Deep In-memory Architectures for Machine Learning

Mingu Kang • Sujan Gonugondla
Naresh R. Shanbhag

Deep In-memory Architectures for Machine Learning

 Springer

Mingu Kang
IBM Thomas J. Watson Research Center
Yorktown Heights, NY, USA

Sujan Gonugondla
University of Illinois at Urbana-Champaign
Urbana, IL, USA

Naresh R. Shanbhag
University of Illinois at Urbana-Champaign
Urbana, IL, USA

ISBN 978-3-030-35973-7   ISBN 978-3-030-35971-3   (eBook)
https://doi.org/10.1007/978-3-030-35971-3

This Springer imprint is published by the registered company Springer Nature Switzerland AG.
The registered company address is: Gewerbestrasse 11, 6330 Cham, Switzerland

Preface

The concept of deep in-memory architecture (DIMA) described in this book is based on the doctoral dissertation research of the first two coauthors conducted under the supervision of third coauthor. In 2013–2014, we were in search of a circuit fabric that would exhibit a favorable energy vs. SNR trade-off. The rationale underlying our search was the formulation of a Shannon-inspired model of computation, elements of which we had developed gradually over the past decade and which could exploit such a trade-off to realize machine learning accelerators operating at the fundamental limits of energy-latency-accuracy. By embedding analog computations in the periphery of the memory bitcell array, we were not only able to realize such a circuit fabric but also overcome the memory-processor bottleneck inherent in traditional von Neumann digital architectures.

Developing DIMA was a challenging prospect because it broke many traditional assumptions made in the design of circuits, architectures, and algorithms. The justification for breaking those assumptions required us to connect across the compute layers, thereby making DIMA a full-stack technology. This cross-layer connection is hard to comprehend given the siloed nature of modern-day research enterprise. Not surprisingly, we have had our fair share of rejections of our papers by top conferences when we first started. For example, though our first DIMA paper was published in 2014, it was not until 2018 that the general idea of in-memory computing caught the imagination of circuit designers and architects.

Our journey leading to this book would not have been possible without the support and contributions of a number of individuals and institutions. We begin by acknowledging the numerous discussions and brainstorming sessions with Ameya Patil, Charbel Sakr, Sungmin Lim, and Yongjune Kim, who as members of the Shanbhag Research Group at the University of Illinois at Urbana-Champaign directly influenced many of the ideas described in this book. We appreciate the valuable guidance and feedback from Professors Naveen Verma, Pavan Hanumolu, Boris Murmann, and Rob Rutenbar. Their insightful comments and wisdom helped nurture our then nascent concept of deep in-memory architecture (DIMA) into its current form. Specifically, collaborations with Professor Naveen Verma and his students played a critical role in developing DIMA in its various forms. The material in Chap. 6 is based on a highly productive research collaboration with Prakalp Srivastava, and Professors Vikram Adve and Nam-Sung Kim, who introduced us to the intricacies of programming languages, instruction set architecture, and compiler design.

Our work was initiated under the auspices of the Systems on Nanoscale Information fabriCs (SONIC) Center (2013–2017) at the University of Illinois at Urbana-Champaign (UIUC). We feel extremely fortunate to have been part of the SONIC community—its faculty, students, and sponsors—which provided us with a fertile and vibrant environment that inspired, encouraged us to explore high-risk high-payoff ideas such as DIMA, and supported our efforts throughout the process. We gratefully acknowledge the contributions of Sean Eilert and Ken Curewitz from Micron Inc., who provided encouragement and feedback on our work specially in its early days. Our collaborators from Intel Corporation, Sasi Manipatruni, Dmitri Nikonov, and Ian Young, introduced us to challenging long-term problems which both excited and challenged our creativity. We appreciate their proactive involvement in our work and their contributions to SONIC's overall success.

Our work would not have been possible without the generous and sustained sponsorship from Linton Salmon (DARPA) and Todd Younkin (SRC—Semiconductor Research Corporation). To this day, both DARPA and SRC are playing a critical role in promoting advanced semiconductor research in the United States and thereby enabling the USA to maintain its leadership in this field.

Our work was conducted at the Coordinated Science Laboratory and the Department of Electrical Engineering at the University of Illinois at Urbana-Champaign. We gratefully acknowledge access to their world-class facilities from office space, to computing and laboratory infrastructure, to administrative and editorial support that made this book possible. We thank Jenny Applequist for her careful reading of our manuscript which improved its quality immensely.

Finally, we would like to acknowledge the critical role played by the members of our respective families: (first author) Steven, Brandon, Stella, and Yiseul; (second author) Surya, Sunitha, and Meghana; (third author) Vinay, Hailan, and Lei. This book would not have been possible without their strong and sustained support that saw us through many difficulties along the way.

Old Tappan, NJ, USA   Mingu Kang
Urbana, IL, USA   Sujan Gonugondla
Urbana, IL, USA   Naresh R. Shanbhag

The original version of this book was revised: Affiliation of the Author "Mingu Kang" has been updated. A correction to this book is available at https://doi.org/10.1007/978-3-030-35971-3_8

Contents

Chapter 1
Introduction

There is much interest in incorporating artificial intelligence (AI) capabilities into various products and services in both the commercial and defense industries today. Though machine learning (ML) algorithms have begun to exceed human performance in cognitive and decision-making tasks [1, 2], they tend to be computationally complex and require processing of large volumes of data. Today, such tasks are realized in the Cloud [3] (see Fig. 1.1a) due to the availability of sufficient computational resources. However, there is growing interest in embedding data analytics into sensor-rich platforms at the Edge including wearables, autonomous vehicles, personal biomedical devices, Internet of Things (IoT) devices and others to provide them with local decision-making capabilities as shown in Fig. 1.1b. Such platforms, though (sensory) data-rich, are heavily constrained in terms of computational resources (storage, processing, and communications), energy, latency, and form factor. Therefore, there is much interest in exploring the implementation of machine learning systems on resource-constrained Edge platforms. This book describes a unique approach to realizing that objective—the *Deep In-memory Architecture* (DIMA).

This chapter begins with description of the energy problem in realizing machine learning workloads on the traditional (digital) von Neumann architecture, followed by a comparative overview of various architectures for implementing machine learning systems.

1.1 The Energy Problem in Machine Learning

Primary metrics for the design of machine learning systems on resource-constrained Edge platforms are: (1) energy-efficiency; (2) decision latency and throughput; and (3) decision(-making) accuracy. Energy efficiency is critical for embedded battery-

© Springer Nature Switzerland AG 2020
M. Kang et al., *Deep In-memory Architectures for Machine Learning*,
https://doi.org/10.1007/978-3-030-35971-3_1

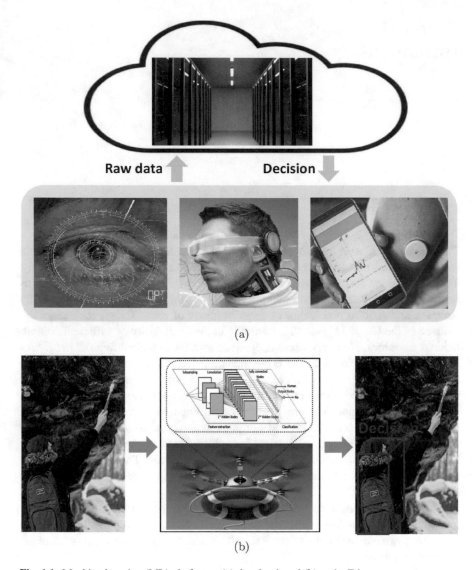

Fig. 1.1 Machine learning (ML) platforms: (**a**) the cloud, and (**b**) at the Edge

powered and autonomous platforms. As a result, a number of integrated circuit (IC) implementations of ML kernels and algorithms have appeared recently [4–9].

For the implementation of ML systems, the mainstream architecture today is the digital (von Neumann) architecture Fig. 1.2a. A key feature of this architecture is the intrinsic separation between memory and processor, i.e., the memory-processor interface. The energy and latency of an ML system realized on a digital architecture include the energy and latency costs of memory accesses via the memory interface and those of the arithmetic operations executed in the processor. It has been well-

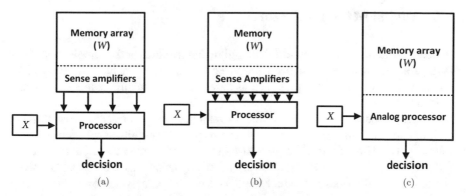

Fig. 1.2 Inference architectures: (**a**) digital (von Neumann) architecture, (**b**) near-memory, architecture, and (**c**) in-memory architecture, where X is an input pattern and W is a stored weight parameter

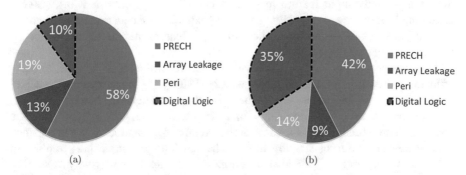

Fig. 1.3 Energy breakdown for two simple pattern matching tasks in a 65 nm process using (**a**) Manhattan distance, and (**b**) cross correlation to find the closest image to a query from 64 candidate images. A 512×256 SRAM with synthesized 8-b digital logic is simulated post-layout. The dotted region is from the digital computational block including the clock tree network. The peripheral circuitry (Peri) includes the energy of the sense amplifiers (SAs), decoders, 4:1 column muxes, and WL drivers

established that, the energy and latency costs are dominated by memory accesses [10], e.g., it takes about 20–100 pJ per access of a 16-b word from a 32 kB to 1 MB static random access memory (SRAM) versus 1 pJ per multiply in a 45 nm process [10]. As an additional confirmation of this energy gap, the energy breakdown for two pattern matching tasks realized in a 65 nm process is shown in Fig. 1.3, which shows that memory access energy dominates by taking up to 90% of the total system energy. In addition, recent implementations of deep neural networks (DNN) [4, 9] also report that memory accesses account for the largest portion (between 35% and 45%) of the total energy cost.

Therefore, it is clear that one needs to address the high energy costs of memory accesses even if at the expense of computational energy, if AI capabilities are to be realized on Edge devices.

1.2   Digital ML Architectures

Recently, a number of energy-efficient digital architectures and IC implementations [4–9] for ML systems have been proposed that strive to reduce the number of memory accesses via techniques such as data reuse, minimizing computations, and efficient data-flow.

Several research efforts have tried to minimize the data access cost through architectural optimization. Processor-in-memory (PIM) architectures such as Smart Memory [11–13] and Intelligent RAM [14] locate frequently used logic (e.g., pointer logic [13] or MAC [14]) close to memory by using a wide crossbar. However, physical proximity does not reduce memory read and processing costs.

An effective approach in reducing memory access energy and enhancing through-put for ML algorithms is to reuse data once read from memory. DianNao [9] first identified and exploited the opportunity for massive data reuse across the fetched tile of input and output feature maps in a convolutional neural network (CNN), achieving 21× energy savings. Eyeriss [4, 15] extended the data reuse opportunities at multiple levels (convolutional, filter, and input feature map reuse) to achieve up to 2.5× energy savings for AlexNet.

Low-power circuit techniques have also been explored to achieve energy-efficient processing for ML algorithms. ENVISION [5] implemented the CNN with a dynamic voltage-accuracy-frequency scaling technique given a bit-precision requirement. A speech recognizer with a deep neural network (DNN) [16] has also been introduced that has a voice-activated power-gating technique to enhance energy efficiency during standby mode. The RAZOR technique [17] applied to a sparse DNN engine [7] allows minimizing the supply voltage by tolerating timing errors. These approaches achieve significant energy efficiency, but without exploiting the opportunities afforded by analog processing.

Low-voltage SRAM techniques have been proposed [18, 19] to reduce the energy of memory read accesses. These techniques involve operating the bitcell array (BCA) at voltages in the range of a few hundred mVs, which reduces the throughput significantly into the kHz regime. Low-voltage operation also degrades SRAM's read and write static margins causing catastrophic failure of inference applications when MSB errors occur. Therefore, SRAMs tailored for inference algorithms have been proposed as in [20, 21]. In these works, selective bitline (BL) negative boosting is employed to improve write-ability and protect MSBs during the read operation by the means of selective error correcting code (ECC). However, those techniques suffer from large BL-toggling energy and dropping out of LSBs to accommodate ECC check bits. In [22], a filter approximation technique was employed and accelerated by a 7T SRAM to fetch convolution filter coefficients efficiently by enabling two read modes: row-access and column-access, but at the cost of degraded storage density due to the use of an additional transistor in the bitcell.

To sum up, previous approaches have addressed the energy cost by either co-locating the processor and memory, or by minimizing data accesses (via data reuse) or by employing low-power digital techniques for processing and low-voltage memories. In contrast, with associative memories [23, 24], simple logic operations are embedded into the BCA to determine a data vector with the minimum Hamming or Euclidean distances from a reference data vector. This embedding is done at the expense of storage density due to the presence of logic circuits in the bitcell. *Kerneltron* [25] also embeds computation (bit-wise multiplication) into the BCA to process and read simultaneously in the charge domain. However, this requires the use of charge injection devices in the BCA and a massive array of ADCs to interface analog and digital processing. Moreover, the need for special devices makes the Kerneltron incompatible with mainstream memory topologies such as SRAM or DRAM.

Near-memory architectures [26–28] (see Fig. 1.2b) distribute computations near memory banks or replace the digital processor with an analog processor in order to save both computational and memory access energy. However, the near-memory architectures preserve the intrinsic separation between the memory and processor.

1.3 In-Memory ML Architectures

Recently, in-memory architectures [29–37] were proposed in which memory access costs in ML systems have been directly addressed by embedding analog computations in close proximity to the memory bitcell array (BCA). Such proximity of computation to the BCA makes in-memory architectures inherently and massively parallel, and well-matched to the data-flow of ML algorithms. However, their intrinsic analog nature makes in-memory architectures susceptible to process, voltage, and temperature (PVT) variations. The use of BL swings and stringent bitcell pitch-matching constraints results in lowered signal-to-noise (SNR) of in-memory computations. Those robustness issues were addressed by a combination of (1) restricting one or both operands to be binary (1-b) [35, 36], (2) using larger (8T or 10T) bitcells [34], and (3) partitioning the array into sub-banks [34] or using separate right and left WLs [36]. Specifically, the restriction on 1-b operands makes it difficult for such in-memory architectures to execute multi-bit operations without sacrificing their energy-latency benefits. Furthermore, the use of binary nets [38], the default network realized by these precision restricted in-memory architectures, leads to increased memory requirements for a fixed accuracy as shown theoretically in [39] and experimentally in [40].

In contrast, the deep in-memory architecture (DIMA) [29–33, 41, 42] embeds *multi-bit mixed-signal computations* in the periphery of a conventional 6T BCA to preserve its storage density, and conventional SRAM read-write functionality. DIMA accesses multiple rows of a standard SRAM BCA per precharge to generate a BL discharge ΔV_{BL} that is proportional to a linear combination of column bits. DIMA then performs scalar operations on the BL outputs via column pitch-matched mixed-signal circuitry in the periphery of the SRAM, followed by a step that aggregates the BL outputs across all the columns of the BCA. Thus, DIMA computations are intrinsically multi-bit, e.g., 8-b in [32, 33] and 5-b in [31], and strives, to delay the conversion from analog to digital to allow for SNR budgeting across the different stages of analog processing. It can be shown that this *delayed decision property* of DIMA, combined with its cross BL aggregation step effectively compensates for the low-SNR problem inherent to in-memory architectures and thereby leads to accurate inferences [32]. Algorithmic approaches, such as boosting (e.g., AdaBoost) [31] and ensemble methods (e.g., random forest (RF)) [33, 42], have also been leveraged to enhance DIMA's robustness to PVT variations.

1.4 Book Organization

This book is organized as follows. Chapter 2 provides an overview of DIMA. Since DIMA is analog, the chapter provides a systems rationale to quantify the intrinsic robustness of in-memory architectures to analog non-idealities. The rest of the chapter (and the book) focuses on DIMA, since it represents the most aggressive form of in-memory architectures. DIMA design techniques and guidelines are provided to address the circuit and architectural implementation challenges. In addition, this chapter provides energy, delay, and behavioral models with key design parameters.

Chapter 3 presents two DIMA prototype ICs in a 65 nm CMOS process: (1) a multi-functional DIMA that realizes four algorithms, and (2) the random forest (RF) DIMA IC. This chapter also describes techniques for achieving reconfigurability with analog circuitry.

Chapter 4 presents a DIMA IC with an on-chip training loop whose intent is to push the limits of decision-level EDP gains by adapting the model parameters to variations in process parameters, temperature, and voltage as well as data statistics. This chapter shows that to operate at the limits, some form of adaptation and error compensation is required.

While the earlier chapters focus on realization of simple ML kernels based on efficient matrix-vector multiplication (MVM) on DIMA, Chap. 5 describes the mapping of complex ML algorithms such as deep neural networks (DNNs), convolutional neural networks (CNN), and sparse distributed memory (SDM). In particular, algorithmic reformulations such as error-aware retraining (for CNNs) and hierarchical decision-making (for SDM) are shown to enhance the benefits of using DIMA.

Chapter 6 describes a DIMA-based accelerator core with an instruction set architecture (ISA) that is aligned well with the common functional flow of multiple ML benchmarks.

Chapter 7 concludes this book and provides perspectives on future work and trends.

Chapter 2
The Deep In-Memory Architecture (DIMA)

This chapter describes the Deep In-memory Architecture (DIMA). First, the algorithmic data-flow of commonly used ML algorithms is described. DIMA's architectural data-flow is shown to be well-matched to the algorithmic data-flow of those algorithms since it implements a highly efficient matrix-vector multiply (MVM) operation. The rationale underlying DIMA's robustness to PVT variations is presented next. The various stages of analog computations in DIMA are described in detail along with practical circuit and architectural design guidelines. Finally, circuit-aware system models of DIMA's energy, delay, and functionality are presented and employed to evaluate DIMA's decision-making accuracy, energy, and latency trade-offs.

2.1 Data-Flow of Machine Learning Algorithms

Current-day machine learning algorithms require the implementation of massive numbers of dot-products (DPs). A DP is a measure of the Euclidean distance between two N-dimensional vectors, W and X, and is typically followed by a non-linearity as shown below:

$$y = \sum_{i=1}^{N} \mathbf{w}^\mathsf{T}\mathbf{x} \tag{2.1}$$

$$\tilde{y} = f(y) \tag{2.2}$$

where y is the DP output (a scalar), $\mathbf{w} = [W_1, W_2, \ldots, W_N]^\mathsf{T}$ and $\mathbf{x} = [X_1, X_2, \ldots, X_N]^\mathsf{T}$, and $f(\ )$ is a non-linear function, e.g., a rectified linear unit (ReLU) or sigmoid. In the context of DIMA, the $\mathbf{w}$ is the stored weight vector and $\mathbf{x}$ is the input vector. The algorithmic data flow of (2.1) in Fig. 2.1a shows that the

© Springer Nature Switzerland AG 2020
M. Kang et al., *Deep In-memory Architectures for Machine Learning*,
https://doi.org/10.1007/978-3-030-35971-3_2

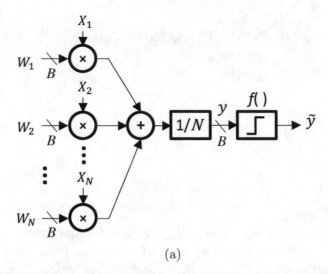

(a)

$f\left(D\left(\mathbf{w}, \mathbf{x}\right)\right)$	Inner loop kernel $D(\mathbf{w}, \mathbf{x}) = \sum_{i=1}^{N} d(W_i, X_i)$	$f()$
SVM	$\sum_{i=1}^{N} W_i X_i$	sign
Temp. Match. (L1)	$\sum_{i=1}^{N} |W_i - X_i|$	min
Temp. Match. (L2)	$\sum_{i=1}^{N} (W_i - X_i)^2$	min
DNN	$\sum_{i=1}^{N} W_i X_i$	sigmoid
Feature extraction (PCA)	$\sum_{i=1}^{N} W_i X_i$	—
k-NN (L1)	$\sum_{i=1}^{N} |W_i - X_i|$	majority vote
k-NN (L2)	$\sum_{i=1}^{N} (W_i - X_i)^2$	majority vote
Matched filter	$\sum_{i=1}^{N} W_i X_i$	min
Linear Regression	$\sum_{i=1}^{N} W_i$	accumulate
	$\sum_{i=1}^{N} W_i^2$	accumulate
	$\sum_{i=1}^{N} W_i X_i$	accumulate

(b)

Fig. 2.1 Typical inference algorithms: (**a**) algorithmic data-flow, and (**b**) specifics of a set of commonly used algorithms

DP is obtained by aggregating element-wise scalar products. ML algorithms may employ other notions of VD, as shown in Fig. 2.1b.

ML algorithms have inherent error resiliency for the following reasons: (1) the final thresholding operation to obtain class labels is immune to small-magnitude errors at its input, and (2) the aggregation of a large number of elements in computing a VD enhances the SNR if the computational noise in the constituent SDs is uncorrelated. Those sources of the intrinsic error-tolerance of ML algorithms can be exploited by both the digital architecture and DIMA to reduce the energy and throughput costs of inference. However, as shown later, DIMA is able to exploit the algorithmic error-tolerance much more effectively.

2.2 DIMA Overview

A high-level view of DIMA (Fig. 2.2a) [29, 41, 43, 44] shows that the data stored in an $N_{ROW} \times N_{COL}$ bitcell array (BCA) are processed in four sequentially executed stages:

- the **multi-row functional read** (FR) stage fetches an N-dimensional data vector W consisting of column-major stored B-bit elements by reading B rows (*word-row*) per BL precharge (read cycle). This stage includes the precharge circuitry, the FR WL drivers and the BCA;
- the **BL processing** (BLP) stage computes scalar distances (SDs) between the elements of W and X. The N_{COL} BLP blocks operate in parallel in a single-instruction multiple-data (SIMD) manner;
- the **cross BL processing** (CBLP) stage aggregates the SDs computed by the N_{COL} analog BLP blocks to generate a vector distance (VD); and
- in the final stage an **analog-to-digital converter** (ADC) and **residual digital logic** (RDL) are used to realize for realizing any digital computations that may be necessary such as thresholding/decision function $f()$ and others.

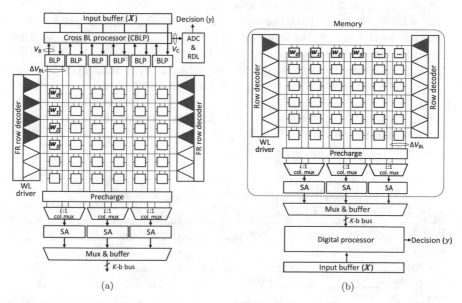

Fig. 2.2 Inference architectures: (**a**) DIMA, and (**b**) the digital (von Neumann) architecture. The blocks marked in red are the wordline drivers that are turned on

Thus, the final output of DIMA from one read cycle is a scalar y representing the result of a DP operation in (2.1). The scalar non-linearity in (2.2) can be implemented either digitally, or by clipping the ADC outputs or setting the ADC thresholds appropriately.

A comparison of DIMA (Fig. 2.2a) and a von Neumann (digital) architecture (Fig. 2.2b) shows that both employ identical BCAs to store W and an input buffer to store streamed X. However, in DIMA, N_{COL} SDs are computed via BLPs in analog right next to the BCA, while the digital architecture needs to access data fully and accurately before processing.

This is because conventional SRAMs need to employ column multiplexing, whereby multiple BLs share a single sense amplifier (SA). Reliability constraints force the SA and other peripheral circuits to be designed with area that is $4\times$ to $8\times$ of that of a bitcell, thereby necessitating column multiplexing. In addition, there is another throughput bottleneck due to the limited memory I/O port or bus width in von Neumann architectures [45]. Typically, L read outs are required in order to read all of the data in a single row, even in application processors with custom-designed on-chip SRAM [46].

In contrast, DIMA is able to bypass the column muxing requirements imposed on conventional SRAMs. Furthermore, DIMA can directly aggregate the outputs of BLPs with CBLP to generate a VD. Thus, the final output of the analog section in DIMA is a VD instead of data bits as in the case of a digital architecture. The differences are summarized in Table 2.1 and elaborated below.

- **Storage pattern**: DIMA stores B bits of W in a column-major format vs. row-major used in digital architectures (Fig. 2.2a).
- **Read access**: For each BL precharge (read cycle), DIMA reads a function of B rows or a *word-row*, rather than a single row as in the digital architecture. That process, referred to as the *multi-row functional read* (FR), generates a BL voltage drop ΔV_{BL} proportional to a weighted sum of the B bits per column [29] by using pulse width modulated (PWM) (Fig. 2.3b) or pulse amplitude modulated (PAM) WL signals [31, 47]. Thus, DIMA needs much fewer precharge cycles to read the same number of bits, which leads to both energy and throughput gains. However, in exchange, DIMA relaxes the fidelity of its reads as long as it falls within the error tolerance of the ML algorithm.

Table 2.1 DIMA vs. digital architecture (with $N_{COL} \times N_{ROW}$ bitcell array)

Attribute	Conventional	DIMA
Data storage pattern	Row major	Column major
Column mux ratio	L:1	1:1
Fetched words per access	$N_{COL}/(LB)$	N_{COL}
BL swing/LSB (ΔV_{BL})	250–300 mV	5–30 mV
# of rows per access	1	B
WL driver	Fixed pulse width	Pulse width/amp modulated

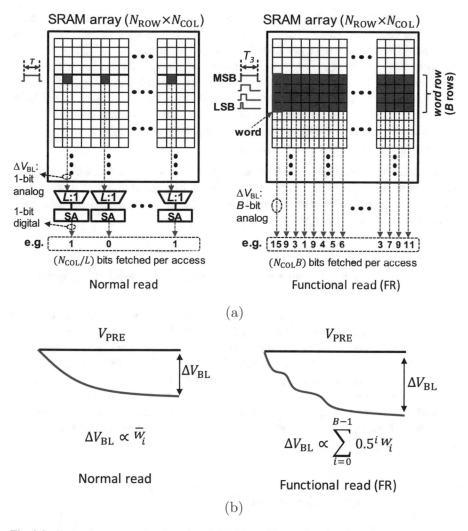

Fig. 2.3 Comparing conventional read and DIMA multi-row functional read (FR) operations: (a) fetched data, where $B = 4$ and $L = 4$ assumed (red marked bitcells read simultaneously), and (b) BL swing (ΔV_{BL})

- **Column muxing**: Unlike standard SRAMs, which require an $L : 1$ column mux ratio (typically $L = 4$–32) to accommodate a large-area sense amplifier (SA) as shown in Fig. 2.2a, DIMA bypasses the muxing via SD computations in BLPs whose horizontal dimension is matched to the column pitch (i.e., the *pitch-matching constraint*) of the BCA. Column muxing limits the number of bits per access to N_{COL}/L in standard SRAM opposed to $N_{COL}B$ in FR.
- **Data reduction and decision**: While the conventional architecture computes in a digital processor, DIMA implements SD and VD via BLP and CBLP right next

to the BCA by using charge-based analog circuits. An ADC is used to digitize the analog CBLP output and pass it on to the RDL to compute $f()$ as in Fig. 2.2b. That ADC operates once per 128–256 SD computations.

- **Accuracy vs. energy**: DIMA computations necessarily have a lower signal-to-noise ratio (SNR) than the computations in the digital architecture. This loss in SNR arises from the spatial transistor threshold voltage variations in the BCA (which affect the FR process), and from severe area-constraints on the BLP and CBLP. However, the SNR can be tuned to a level required by the ML algorithm, e.g., by adjusting the BL swing ΔV_{BL}.

In summary, the key to DIMA's speed-up and energy advantages over a digital architecture arises from its ability to read $N_{\mathrm{COL}} B$ bits per access by FR, bypassing the column muxing requirements, and employing low-swing analog computations, but at the expense of having a lower SNR in computation.

2.3   Inference Architectures: A Shannon-Inspired Perspective

A natural concern with DIMA is its impact on the accuracy of its decisions, given its intrinsic analog nature. In practice, DIMA exhibits an intrinsic robustness to PVT variations due to its high-dimensional computations. This section employs a Shannon-inspired perspective to contrast DIMA and a digital architecture in terms of their robustness. In this perspective, the process of reading data from memory is viewed as a transfer of the data over a (noisy) communication channel thereby making it possible to employ a Shannon-theoretic interpretation of memory read in order to comprehend the robustness of computation, i.e., by amortizing the cost of a read cycle across a large number of computations.

2.3.1   Inference Architectures

Figure 2.4 shows a Shannon-inspired perspective of the digital architecture and DIMA. In this perspective, the DP computation followed by non-linearity is viewed as a process of preserving the information content in data as it progresses from memory to the final decision point.

Digital architectures can exploit the intrinsic error-tolerance of ML algorithms by allowing errors to occur either during the memory read or during computation. In both cases, though, such errors turn out to be catastrophic. For example, *near threshold voltage* (NTV) operation and reduced BL swing in a SRAM can reduce the energy consumption of digital computation and memory accesses, respectively. However, in both cases, there is a heavy loss in inference accuracy since these errors can occur in the most significant bits (MSBs).

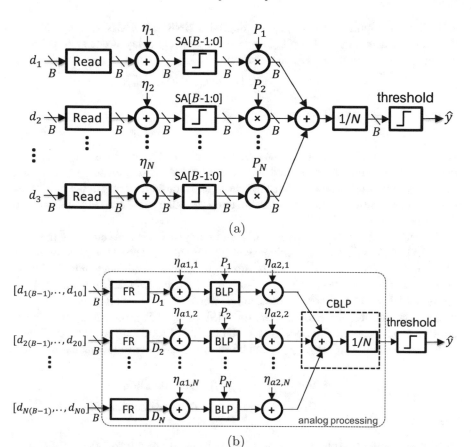

Fig. 2.4 A Shannon-inspired view of inference architectures: (**a**) conventional, and (**b**) DIMA

This is because the BL discharge ΔV_{BL} needs to be converted to full swing to be processed in digital logic through an SA. The SA makes bit-wise hard decisions (Fig. 2.4a) before the subsequent DP computation. Thus, decision errors can occur early in the MSBs due to the noise sources $\eta_1, \eta_2, \ldots, \eta_N$, and as ΔV_{BL} is reduced further, the error magnitude can increase beyond the application's error-tolerance, leading to a catastrophic increase in error-rates of the inference system. For example, it has been shown that the peak signal-to-noise ratio (PSNR) can drop from 50 dB to below 10 dB in the presence of MSB errors [20, 21]). To address this issue, several low-power SRAM techniques [20, 21] have been proposed based on selective MSB protection to achieve moderate (35%) energy savings in memory read accesses. In addition, NTV also leads to a significant ($\approx 10\times$ [48]) delay penalty and increased sensitivity to PVT variations.

In contrast, DIMA budgets its compute SNR by employing low-swing computations in both memory and computation (see Fig. 2.4b). It does so by employing FR, which implicitly performs D/A conversion, and subsequent mixed-signal BLP and CBLP computations. Despite its degraded SNR, DIMA is able to preserve its inference accuracy by adopting three principles: (1) delayed decision, (2) non-uniform bit protection, and (3) aggregation.

2.3.2   Delayed Decision

Because of the elimination of SAs between memory and processor in DIMA, the loss of information associated with early hard decisions made right after BL discharge is avoided. The only hard decision (thresholding) occurs in the final stage (Fig. 2.4b) in contrast to a digital architecture. In other words, DIMA delays hard decisions until the very end.

To see the difference between early and delayed decisions, consider the case (Fig. 2.5) of binary decision making in which a input $x \in \{1, -1\}$ with equal priors $P(X = 1) = P(X = -1) = 0.5$ is subject to identical and independent additive noise sources $\eta_1, \eta_2 \sim \mathcal{N}(0, \sigma_n^2)$. For early decision (Fig. 2.5a), a threshold is applied after each noise source addition, as would be the case when an SA is used in a digital architecture. In contrast, for DIMA (Fig. 2.5a), one final threshold is applied to generate the final decision $\hat{x}$. All threshold levels are set at zero. The bit error probabilities $p_e = \Pr\{x \neq \hat{x}\}$ can be shown to have the following expressions:

$$
p_e = \begin{cases} 2Q(\frac{1}{\sigma_n})[1 - Q(\frac{1}{\sigma_n})] & \text{early decision} \\ Q(\frac{1}{\sqrt{2}\sigma_n}) & \text{delayed decision} \end{cases} \tag{2.3}
$$

Figure 2.5b plots (2.3) to show that delayed decision has a lower p_e, and hence is more accurate, than early decision, but at lower SNRs. This result clearly shows the importance to preserving information by avoiding hard decisions early in the processing chain. For SNR-constrained scenarios such as DIMA, the delayed decision-making is critical in preserving accuracy.

2.3.3   Significance-Based Swing Allocation

Functional read with PWAM (Fig. 2.2b) allows us to assign BL voltage swing based on the significance of the bit, i.e., the bit position, with MSBs allocated more swing than LSBs. When the total swing ΔV_{BL} fixed when reading B bits on a BL, the swing assigned for the nth bit position ΔV_n is given by:

$$
\Delta V_n = \begin{cases} \dfrac{\Delta V_{BL}}{B} & \text{(significance-based)} \\ \dfrac{\Delta V_{BL} 2^{n-1}}{(2^B - 1)} & \text{(conventional)} \end{cases} \tag{2.4}
$$

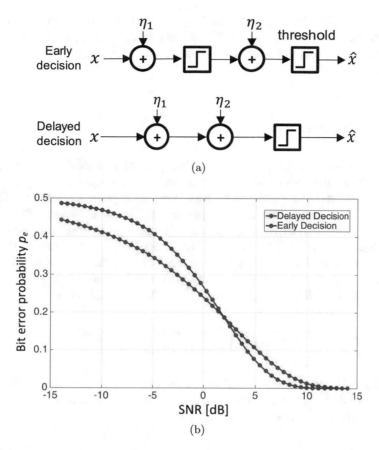

Fig. 2.5 Delayed vs. early decision scenarios: (**a**) simplified functional flow, and (**b**) bit error probability p_e vs. SNR

For example, significance-based swing allocation assigns approximately $2\times$ higher swing to MSB and $3.8\times$ less swing for LSB compared to a uniform swing assignment when $B = 4$. In that way, DIMA efficiently budgets a limited resource (the BL voltage swing range) to minimize the impact of errors on system accuracy.

2.3.4   Massive Aggregation

The charge-sharing process in the CBLP (Fig. 2.4b) averages out the noise contributions across the N BLs to scale the standard deviation of the noise seen at its output by a factor of $1/\sqrt{N}$. That averaging assumes that the N noise sources are independent, zero-mean random variables, which is a reasonable assumption to make.

Since the CBLP averages out both η_1 and η_2 before making a hard decision (thresholding), it results in an SNR boosting effect before the final decision. That effect is not present in a digital architecture which aggregates after the SA hard decision.

2.3.5 Measured Results

This section validates the effectiveness of the principles underlying DIMA's robustness to PVT variations as described in Sects. 2.3.2, 2.3.3, and 2.3.4 by employing measured results of the prototype IC in a 65 nm CMOS process described in Chap. 3.

Figure 2.6a shows that the bit error-rate (BER) of a standard SRAM read operation, measured by fetching 16 KB as a function of ΔV_{BL}, increases rapidly for $\Delta V_{BL} < 230$ mV. The measured BER from Fig. 2.6a is injected into system simulations to evaluate the impact on the accuracy of face detection using SVM algorithm. Figure 2.6b plots the probability of detection (accuracy) as a function of the average BL voltage swing assigned per bit $\Delta V_{BL,avg}$ defined as:

$$\Delta V_{BL,avg} = \begin{cases} \Delta V_{BL} & \text{(digital architecture)} \\ \frac{\Delta V_{BL}}{B} & \text{(DIMA)} \end{cases}$$

in order to enable a fair comparison of the two architectures since DIMA's FR fetches B-bits within a BL swing ΔV_{BL}, in contrast to the digital architecture, which reads a single bit in the same swing.

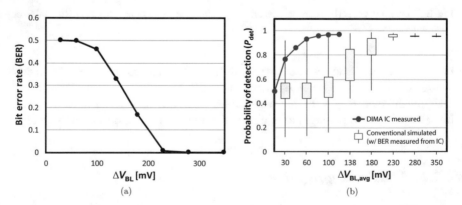

(a)   (b)

Fig. 2.6 Impact of BL voltage swing reduction on the SA bit error-rate (BER) read and system accuracy for face detection with SVM: (**a**) BER of standard SRAM read operation measured from the prototype IC [32], and (**b**) a comparison of system accuracy of the digital architecture (obtained by simulating a digital architecture with measured SRAM read BER) and DIMA (measured from prototype IC [32])

The setup for SVM, including the dataset and image size, is described in Chap. 3. It can be seen that a BER= 7×10^{-3} at $\Delta V_{BL} = 230$ mV can cause 4% degradation in the detection accuracy of the digital architecture (see Fig. 2.6b). In contrast, DIMA's measured accuracy is similar to that of the digital architecture for $\Delta V_{BL,avg} > 100$ mV and robust to lowering of $\Delta V_{BL,avg}$ due to the above mentioned three principles. In fact, the effective ΔV_{BL} of the digital architecture is L times greater than the one shown in Fig. 2.6b as L BLs need to be discharged to access one BL due to column muxing. Therefore, one can conclude that DIMA permits more aggressive ΔV_{BL} scaling than the digital architecture does, leading to significant energy savings.

2.4 DIMA Design Guidelines and Techniques

This section describes the principle of operation of each of the DIMA processing stages and presents design guidelines and techniques for minimizing non-idealities that result from DIMA's analog computations.

2.4.1 Functional Read (FR)

The FR stage performs data access and simple SD computations. Two design techniques: (1) sub-ranged read to enhance linearity, and (2) replica bitcells to achieve efficient data-writing capability. Design criteria are presented for choosing key FR parameters, such as the pulse width T_0 and amplitude V_{WL}, in order to minimize the impact of non-ideal behavior from low-swing analog processing.

2.4.1.1 FR Operation

The FR stage generates a BL voltage drop $\Delta V_{BL}(W)$ proportional to the weighted sum $W = \Sigma_{i=0}^{B-1} 2^i w_i$ of the column-major stored data $\{w_0, w_1, \ldots, w_{B-1}\}$ (see Fig. 2.7a). The voltage drop $\Delta V_{BL}(W)$ can be generated via a simultaneous application of PWAM access pulses to multiple rows per precharge cycle. This is in contrast to the use of single-row fixed-width and amplitude pulses for each precharge cycle as in conventional SRAM reads.

Consider FR of B rows using PWM [29] with pulse widths T_i and pulse amplitudes $V_{WL(i)}$ ($i \in [0, B-1]$) as shown in Fig. 2.7b [29]. The charge $\Delta Q_i(w_i)$, drawn from the BL capacitance C_{BL} is given by:

$$\Delta Q_i(w_i) = \overline{w_i} T_i I(T_i) \qquad (2.5)$$

Fig. 2.7 Multi-row functional read (FR) using pulse width modulated access pulses [29]: (**a**) column structure with bitcells, and (**b**) sample waveforms for a 4-bit word ($W = 0000b'$) read-out. The WL pulses, shown as being applied sequentially, can be overlapped in time

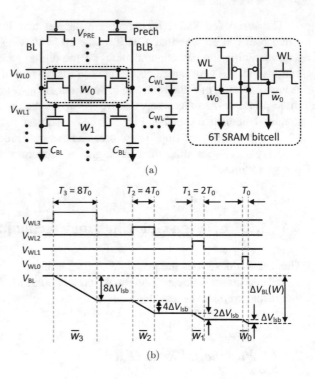

where $I(t)$ is the current drawn by the ith bitcell. This current can be Taylor series approximated as:

$$I(t) = \frac{V_{\text{PRE}}}{R_i} e^{-\frac{t}{R_i C_{\text{BL}}}} \approx \frac{V_{\text{PRE}}}{R_i} \left(1 - \frac{t}{R_i C_{\text{BL}}} \right) \approx \frac{V_{\text{PRE}}}{R_i} \tag{2.6}$$

provided $t \ll R_i C_{\text{BL}}$. Substituting $t = T_i$ into (2.6) and the resulting expression for $I(T_i)$ into (2.5), we obtain:

$$\Delta Q(w_i) = \overline{w_i} T_i \frac{V_{\text{PRE}}}{R_i} \tag{2.7}$$

where $T_i \ll R_i C_{\text{BL}}$. Therefore, the expression for the total BL voltage drop $\Delta V_{\text{BL}}(W)$ can be obtained as follows:

$$\Delta V_{\text{BL}}(W) = \frac{\sum_{i=0}^{B-1} \Delta Q_i}{C_{\text{BL}}} = \frac{V_{\text{PRE}}}{C_{\text{BL}}} \sum_{i=0}^{B-1} \frac{\overline{w_i} T_i}{R_i} \tag{2.8}$$

Equation (2.8) shows that the BL voltage discharge $\Delta V_{\text{BL}}(W)$ is a dot-product between $\{\overline{w_i}\}$ and $\{T_i\}$. If the pulse widths are binary weighted, i.e., $T_i = 2^i T_0$ where T_0 is the LSB pulse width, and if $R_i = R_{\text{BL}}$, i.e., the discharge paths of all the B bitcells in a column have identical resistances, then

$$\Delta V_{\mathrm{BL}}(W) = \frac{V_{\mathrm{PRE}}}{R_{\mathrm{BL}}C_{\mathrm{BL}}} T_0 \sum_{i=0}^{B-1} 2^i \overline{w_i} = \Delta V_{\mathrm{lsb}} \sum_{i=0}^{B-1} 2^i \overline{w_i} = \Delta V_{\mathrm{lsb}} \overline{W} \qquad (2.9)$$

where $\Delta V_{\mathrm{lsb}} = \frac{V_{\mathrm{PRE}} T_0}{R_{\mathrm{BL}} C_{\mathrm{BL}}}$, and $\overline{W}$ is the numerical value of the one's complement of W. The expression in (2.9) is idealized as it assumes the following four conditions:

1. $T_i \ll R_i C_{\mathrm{BL}}$.
2. $T_i = 2^i T_0$.
3. $R_i = R_{\mathrm{BL}}$, i.e., row invariant discharge resistance.
4. R_{BL} is constant over V_{BL}, i.e., bias-invariant discharge resistance.

In practice, those conditions will not be fully met and will lead both to a deviation, i.e., non-linearity, from (2.9), and spatial variations from one group of B bits to another across the BCA. Those non-idealities, and techniques to alleviate them, will be described in Sect. 2.4.1.2.

An expression similar to (2.9) for $\Delta V_{\mathrm{BLB}}(W)$ can be obtained by replacing $\overline{w_i}$ with w_i in (2.9). Thus, the FR stage converts the stored digital data W into BL voltage drops $\Delta V_{\mathrm{BL}}(W)$ and $\Delta V_{\mathrm{BLB}}(W)$, i.e., the FR stage is a coarse digital-to-analog converter. In addition, the FR stage can also realize simple SD functions, such as the addition and subtraction of two B-bit words (W and X) stored in different rows but in the same column. For example, from (2.9), $D + P$ is obtained by applying FR to rows containing W and X to obtain:

$$\Delta V_{\mathrm{BL}}(W + X) = \Delta V_{\mathrm{lsb}} (\overline{W} + \overline{X}) = \sum_{i=0}^{B-1} 2^i (\overline{w_i} + \overline{x_i}) \qquad (2.10)$$

Similarly, subtraction $W - X$ can be realized by storing $\overline{X}$ (one's complement of X) in the same column as W. Subtraction will be discussed in Sect. 2.4.1.3 in more detail.

2.4.1.2 Design Guidelines

The BL swing ΔV_{BL} generated by the FR stage is subject to the impact of spatial transistor threshold voltage variations caused by random dopant fluctuations [49], the voltage-dependence of the discharge path (access and pull-down transistor in the bitcell) resistance R_{BL} (see (2.9)), and the finite transition (rise and fall) times of the PWM WL access pulses. These non-idealities can be incorporated into (2.9) as follows:

$$\Delta V_{\mathrm{BL}}(W) = \Delta V_{\mathrm{lsb}} \sum_{i=0}^{B-1} \frac{2^i \overline{w_i}(1 + \gamma_i)}{(1 + \rho_i(V_{\mathrm{BL}}) + \delta_i)} \qquad (2.11)$$

where δ_i is a random variable describing the impact of spatial transistor threshold voltage variations on the discharge path resistance R_{BL} that affects Condition 3 ($R_i = R_{BL}$) as discussed in Sect. 2.4.1.1, $\rho_i(V_{BL})$ is a variable that captures the impact of the BL voltage dependence of R_{BL} that affects Condition 4 (where R_{BL} should be a constant); and γ_i is a deterministic variable that captures the impact of finite pulse transition times on T_i that affects Condition 2 ($T_i = 2^i T_0$), with the effect on the LSB pulse width T_0 being most severe.

The presence of δ_i, ρ_i, and γ_i imposes certain design constraints that are needed in order to alleviate their impact so that (2.11) approaches the ideal expression in (2.9). For example, it is possible to reduce ρ_i by ensuring that the access transistor in the discharge path does not transit from saturation into the triode region to satisfy Condition 4. That can be achieved by lowering the WL access pulse amplitude V_{WL}, which has the additional benefit that R_{BL} is thereby increased, making it easier to satisfy the overarching Condition 1 ($T_i \ll R_{BL}C_{BL}$). Similarly, the impact of γ_i can be alleviated by ensuring that the design parameter T_0 is lower-bounded as $T_0 > T_{min}$ so that the rise (T_r) and fall (T_f) times of V_{WL} are a small fraction of T_0, e.g., $T_r + T_f < 0.5T_{min}$, of T_0 and hence Condition 2 ($T_i = 2^i T_0$) can be met. That and Condition 1 imply that $T_{min} < T_0 < T_{max}$. Finally, δ_i can be alleviated by ensuring that V_{WL} is sufficiently large so that variations in R_i are reduced, i.e., Condition 3 ($R_i = R_{BL}$) is well-approximated. That lower bound on V_{WL} can be relaxed as DIMA's aggregation process in the CBLP compensates for the impact of δ_i. However, V_{WL} does have an upper bound to avoid destructive read operations, e.g., $V_{WL} < 0.8V_{PRE}$. Hence, for the prototype IC, $V_{WL} = 0.65V_{PRE}$ is chosen. Note that it is possible to pre-distort the data stored in the BCA in order to alleviate the impact of deterministic errors ρ_i and γ_i.

The worst-case values of ρ_i and γ_i are less than 41% and 37%, respectively, as estimated from measured results of the multi-functional 65 nm CMOS prototype IC. Monte Carlo post-layout simulations of the BCA show that the impact of δ_i leads to a 12% variation (σ/μ) in $\Delta V_{BL}(W)$ for typical values of $V_{WL} = 0.65$ V, $V_{PRE} = 1$ V, $T_0 = 250$ ps, $N = 128$, $B = 8$, and $N_{ROW} = 512$. Section 3.2 indicates that those non-idealities have a negligible impact on inference accuracy for the datasets being considered in this book.

2.4.1.3   Design Techniques

We present two design techniques to overcome the design constraints described in Sect. 2.4.1.2.

- *Sub-ranged Read*: Realization of a highly linear FR stage when $B > 4$ bits is challenging because the constraint $T_{min} = 2(T_r + T_f) < T_0 < T_{max} \ll 2^{1-B}R_{BL}C_{BL}$ is hard to meet. For example, $T_0 < 125$ ps when $B = 5$ and $T_4 = 2$ ns. That value of T_0 is hard is achieve when one is driving high WL capacitance (e.g., 200 fF) with a row pitch-matched wordline driver. The sub-ranged read technique solves the problem, as described next.

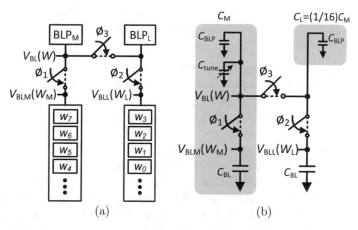

Fig. 2.8 Sub-ranged read with $B = 8$: (**a**) BL pair (two neighboring bitcell columns), and (**b**) equivalent capacitance model [44], where $W_M = 8w_7 + 4w_6 + 2w_5 + w_4$ and $W_L = 8w_3 + 4w_2 + 2w_1 + w_0$

In sub-ranged read [44], the $B/2$ MSBs representing data W_M and the $B/2$ LSBs representing data W_L are stored in adjacent columns of the BCA, as shown in see Fig. 2.8a. For example, when $B = 8$, $W_M = 8w_7 + 4w_6 + 2w_5 + w_4$ and $W_L = 8w_3 + 4w_2 + 2w_1 + w_0$. Three switches $\phi_{1,2,3}$ are used in a specific sequence to charge-share the BL voltages. Explicit tuning capacitor C_{tune} enables the realization of a predefined capacitance ratio (16 : 1 for $B = 8$) between the MSB and LSB BLs' capacitances C_M and C_L, respectively. That ratio is needed to weigh the voltage drop on the MSB BL (ΔV_{BLM}) by a factor of $2^{\frac{B}{2}}$ more than the voltage drop on the LSB BL (ΔV_{BLL}). The desired capacitance ratio is obtained by setting $C_M = C_{BL} + C_{BLP} + C_{tune}$ and $C_L = C_{BLP}$ as shown in Fig. 2.8b, and by varying the C_{tune} modifies C_M so as to realize $C_M : C_L = 2^{\frac{B}{2}} : 1$.

The sub-ranged read proceeds as follows.

1. The FR process is simultaneously applied to the MSB and LSB columns with $\phi_{1,2,3} = 0$ (all open) thereby generating voltage drops $\Delta V_{BLM}(W_M)$ and $\Delta V_{BLL}(W_L)$, respectively.
2. The switch ϕ_1 is closed so that the voltage $\Delta V_{BLM}(W_M)$ is developed across C_M. Switch ϕ_2 is pulsed to generate the voltage $\Delta V_{BLL}(W_L)$ on C_L.
3. The switch ϕ_3 is closed to generate the voltage drop at the final output.

$$\Delta V_{BL}(W) = \frac{1}{C_M + C_L} [C_M \Delta V_{BLM}(W_M) + C_L \Delta V_{BLL}(W_L)] \quad (2.12)$$

$$= \frac{1}{2^{\frac{B}{2}} + 1} \left[2^{\frac{B}{2}} \Delta V_{BLM}(W_M) + \Delta V_{BLL}(W_L) \right] \quad (2.13)$$

Thus, for $B = 8$, the voltage drop $\Delta V_{\text{BLM}}(W_{\text{M}})$ is weighted $16\times$ more than $\Delta V_{\text{BLL}}(W_{\text{L}})$, to obtain the voltage drop $\Delta V_{\text{BL}}(W)$ which is proportional to $16W_{\text{M}} + W_{\text{L}}$.

- *FR Replica BCA*: As described in Sect. 2.4.1.1, two operands W and X are required to implement various SD computations. For example, in order to realize the difference $W - X$ (see (2.10)), $\overline{X}$ is stored in the same column as W but in a different row. Typically, X is a streamed-in data vector, e.g., a template in template matching or image pixels. Storing $\overline{X}$ in the same BCA as W will require repeated SRAM write operations and that would incur large energy and delay costs as these require full BL swing. The problem can be solved via the use of a replica BCA (Fig. 2.9a) which enables fast writes of $\overline{X}$ via separate write BL (WBL) and WL (WWL) [44]. Thus, one can write $\overline{X}$ into the replica BCA column by providing data in a bit-serial manner via the WBL (Fig. 2.9b) while disabling the cross-coupled inverter feedback loop in the replica bitcell via $\overline{\text{WWL}}$. During a subsequent FR, the replica BCA behaves as an extension of the regular BCA. The layout of the replica bitcell needs to be similar to that of a normal bitcell to have the same discharge strength.

2.4.2 BL Processing (BLP) and Cross BLP (CBLP)

This section describes BLP and CBLP stages, which perform various SD computations and aggregation, respectively. The BLP and CBLP operations rely on tightly pitch-matched analog processing (see Fig. 2.10) that result in many implementation challenges. Thus, design principles for key parameters are presented based on the analysis of various noise sources, such as charge injection, thermal noise, and coupling noise.

2.4.2.1   BLP Operation

The N_{COL} BLP block in Fig. 2.2a accepts two operands: (1) its corresponding BL voltage drop $\Delta V_{\text{BL}}(W)$, generated via the FR stage; and (2) a word X to generate an output voltage $V_{\text{B}}(W, X)$. The BLP block needs to be re-configurable in order to support the multiple SD computations required by various ML algorithms. Furthermore, the BLP block layout needs to be column-pitch matched to the BCA. Thus, the BLP stage is a massively parallel analog SIMD processor.

Next, we describe how SD functions such as absolute difference $|W - X|$ [29] and multiplication WX [41] can be implemented in the BLP. These SD functions are required to compute commonly used VDs: Manhattan distance (MD) ($\sum_{i=1}^{N} |W_i - X_i|$) and dot product (DP) ($\sum_{i=1}^{N} W_i X_i$).

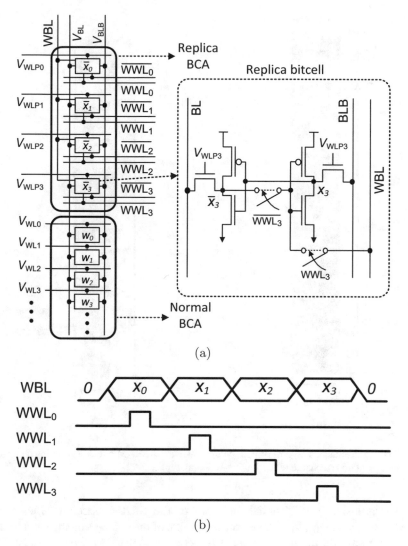

Fig. 2.9 Replica BCA: (**a**) bitcell column ($B = 4$), and (**b**) timing diagram for replica BCA writing [44]

- *Absolute Difference*: The absolute difference $|W - X|$ can be written as [29]:

$$|W - X| = \max(W - X, X - W) \tag{2.14}$$

From (2.10), the BL voltage drop corresponding to the difference $W - X$ is obtained as:

$$\Delta V_{BL}(W - X) = \Delta V_{lsb}(\overline{W} + X) \tag{2.15}$$

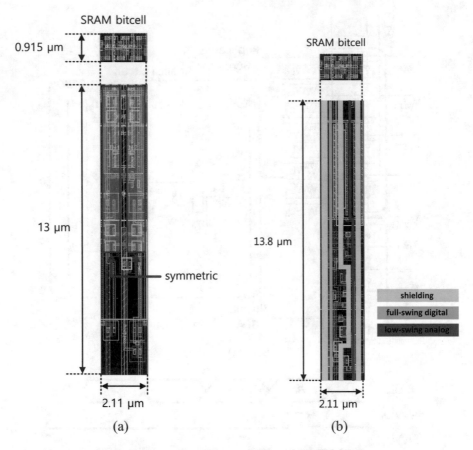

Fig. 2.10 Pitch-matched layouts of BLP blocks relative to a SRAM bitcell: (**a**) analog comparator, and (**b**) a part (1/5) of charge redistribution-based multiplier

The intrinsically differential structure of the SRAM bitcell enables one to evaluate $max(V_{BL}, V_{BLB})$, i.e., the maximum of the voltages on the BL and BLB, quite easily via the use of a local BL compare-select, as shown in Fig. 2.11a. Thus, from (2.14) and (2.15), we get

$$\max(V_{BL}, V_{BLB}) = \max(V_{PRE} - \Delta V_{lsb}(X + \overline{W}), V_{PRE} - \Delta V_{lsb}(W + \overline{X}))$$

$$= \max(V_{PRE} - \Delta V_{lsb}(X + 2^B - 1 - W),$$

$$V_{PRE} - \Delta V_{lsb}(W + 2^B - 1 - X))$$

$$= V_{PRE} - (2^B - 1)\Delta V_{lsb} + \Delta V_{lsb} \max(X - W, W - X)$$

$$(2.16)$$

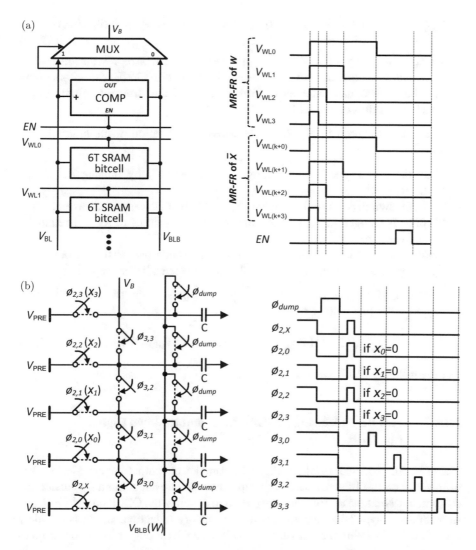

Fig. 2.11 BL processing (BLP): (**a**) absolute difference, where W and $\overline{X}$ are stored in the same column [29], and (**b**) charge redistribution-based multiplication with $B = 4$ [41]

Thus, the simultaneous application of FR to W and $\overline{X}$ results in V_{BL} and V_{BLB} to be proportional to $X - W$ and $W - X$, respectively. The local BL compare-select block (Fig. 2.11a) provides the maximum of V_{BL} and V_{BLB}, and hence the absolute difference $|W - X|$.

- *Multiplication*: Figure 2.11b shows a charge redistribution-based mixed-signal multiplier with inputs $\Delta V_{BLB}(W)$ (FR stage output) and an externally provided B-bit digital word X, whose bits X_i control the $\phi_{2,i}$ switches. The multiplier output voltage V_B is given by:

$$V_B(W, X) = V_B(WX) = V_{PRE} - (0.5)^B X \Delta V_{BLB}(W) = V_{PRE} - (0.5)^B \Delta V_{lsb} W X \tag{2.17}$$

Thus, voltage drop $\Delta V_B(WX) = V_{PRE} - V_B(WX) \propto WX$ represents the product of W and X. Note that the multiplier employs unit size (25 fF) capacitors rather than binary-weighted ones as in [50], because of the stringent column pitch-match constraints on the BLP.

The timing diagram in Fig. 2.11b describes the operation of the multiplier, as follows:

- First, the unit capacitors are charged to $\Delta V_{BLB}(W)$ by pulsing the ϕ_{dump} switches.
- Then, the switch $\phi_{2,i}$ is pulsed only if $X_i = 0$ so that the capacitor corresponding to ith bit is charged to V_{PRE}. The capacitors corresponding to $X_i = 1$ retain the voltage equal to $\Delta V_{BLB}(W)$.
- Finally, the switches $\phi_{3,i}$ are pulsed sequentially starting from $\phi_{3,0}$, $\phi_{3,1} \ldots \phi_{3,B-1}$. That leads to charge sharing between adjacent capacitors.

Note that when $\phi_{3,k}$ ($k = 0, \ldots, B - 1$) is pulsed, the two charge sharing capacitors settle to a voltage of:

$$V_{PRE} - (0.5)^{k+1}(2^k x_k + \ldots + 2x_1 + x_0)\Delta V_{BLB}(W) \tag{2.18}$$

thereby realizing (2.17) when $k = B - 1$.

2.4.2.2 Cross BL Processing (CBLP), ADC, and Residual Digital Logic (RDL)

The cross BL processor (CBLP) (Fig. 2.12) samples the output voltage ($V_B(W, X)$) of the BLP on the BL-wise sampling capacitors C_S at each column by pulsing the ϕ_1 switches. Next the ϕ_2 switches are pulsed to generate the CBLP output V_C in one step. In that way, CBLP implements dimensionality reduction, which is a widely used function in inference algorithms. Finally, the CBLP output V_C is converted to

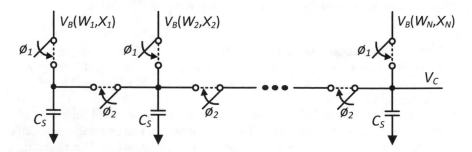

Fig. 2.12 Cross BL processing (CBLP)

the digital domain by the ADC to be stored or further processed by the RDL. The RDL implements slicing/thresholding functions such as min, max, sign, sigmoid, and majority vote. Note that the ADC and RDL need to process one scalar value (V_C) generated from a massively parallel (>128) SD processing step in the BLP. Thus, the energy overhead of ADC and RDL is negligible.

2.4.2.3 Design Guidelines

The BLP can be configured to compute the absolute difference $|W - X|$ (MD mode) or the scalar product WX (DP mode). The dominant source of non-ideality in computing $|W - X|$ is the comparator offset in the compare-select block (see Fig. 2.11a).

However, that input offset affects the BLP output V_B minimally. The reason is that the input offset affects the output only when V_{BL} and V_{BLB} are close to each other. Thus, the BLP output $V_B = max(V_{BL}, V_{BLB})$ in (2.14) is supposed to have an error of only small magnitude $|V_{BL} - V_{BLB}|$. Further, the error in V_B, being uncorrelated across the columns, gets averaged out further by the CBLP. The column pitch-matched comparator layout shown in Fig. 2.10a is constrained to be symmetric to minimize the input offset. Monte Carlo post-layout simulations in the 65 nm CMOS process indicate that the input offset follows the distribution $\mathcal{N}(0, (10\,\text{mV})^2)$.

Computation of the product WX in the BLP (see Fig. 2.11b) and summation in the CBLP (see Fig. 2.12) are done via charge redistribution circuits. These circuits suffer from multiple noise sources: (1) charge-injection noise, (2) coupling noise, and (3) thermal noise. Assuming a junction capacitance of 0.05 fF [51] that uses minimum sized switches, we find that storage capacitance C needs to be larger than 13 fF in order to ensure 8-b output precision. The 8-b precision in a swing of 300 mV results in a resolution of $V_{res} = 1\,\text{mV}$. Hence, thermal noise considerations ($\sqrt{KT/C} < 0.5 V_{res}$) lead to the requirement of $C > 17$ fF at $T = 300$ K. Hence, we chose $C = 25$ fF to provide a sufficient design margin.

Due to the tight pitch-matching constraints, digital signals need to be routed over analog nodes in the BLP and CBLP that generate significant coupling noise. In order to alleviate coupling noise, low-swing analog nodes were shielded from the digital full-swing lines as shown in Fig. 2.10b.

2.5 DIMA Models of Energy, Delay, and Accuracy

This section establishes models and methods that are needed to understand the fundamental energy-delay and accuracy trade-offs underlying DIMA, by (1) presenting silicon-validated energy, delay, and accuracy models; and (2) employing them to identify the most effective design parameters of DIMA to maximize its EDP gains at a given level of accuracy, and to understand the fundamental limits and trade-offs between energy and accuracy.

2.5.1 Energy and Delay Models

This section compares the energy and delay models for the digital architecture and DIMA where we assume that the BCA size $N_{ROW} \times N_{COL}$, the BL precharge voltage V_{PRE}, and the maximum BL swing $\Delta V_{BL,max}$ are identical in both architectures.

Since the SRAM in the digital architecture requires an $L : 1$ column multiplexer (where a typical $L = 4, \ldots, 16$) to accommodate large area sense amplifiers (SAs) as shown in Fig. 2.3a, the number of bits per read cycle is limited to N_{COL}/L compared to $N_{COL}B$ in DIMA's FR. Therefore, DIMA needs LB times fewer read cycles to read the same number of bits. However, the read cycle time for DIMA is larger than that of the digital architecture since DIMA's read cycle includes both data read and compute functions via the FR, BLP, and CBLP stages. Ignoring the compute delay of the digital architecture, we find that the delay reduction factor involved in reading a fixed number of bits is given by:

$$\rho_d = LB \cdot \frac{T_{digital}}{T_{DIMA}} = \frac{LB}{\gamma} \tag{2.19}$$

where $T_{digital}$ and T_{DIMA} are the read cycle times of the digital architecture and DIMA, respectively, and $\gamma = T_{DIMA}/T_{digital} \approx 3$ or 6 (in 65 nm CMOS) depending on the BLP function being computed. Previous work [32] has shown that DIMA can realize $B \leq 6$, with $B = 4$ being comfortably realized. Hence, $\rho_d = 5 \times$ to $21 \times$ is easily achievable with typical values of $B = 4$, $L = 4, 8, 16$ and $\gamma = 3$ with $\rho_d = 5.3$ demonstrated in silicon [32].

The dominant sources of energy consumption in a SRAM array are the dynamic energy consumed to precharge large BL capacitances C_{BL} during every read cycle and because of leakage. The energy consumed in reading B bits in the digital architecture and DIMA can be expressed as:

$$E_{digital} = LBC_{BL}\Delta V_{BL,max}V_{PRE} + E_{lk\text{-}digital} \tag{2.20}$$

$$E_{DIMA} = \beta C_{BL}\Delta V_{BL,max}V_{PRE} + \frac{E_{lk\text{-}digital}}{\rho_d} \tag{2.21}$$

where $E_{lk\text{-}digital}$ is the leakage energy of the digital architecture, and $1 \leq \beta < 2$ is an empirical factor to include DIMA's FR is a pure B-bit read or incorporates a $2B$-bit computation in the FR stage [32]. The leakage energy of DIMA is reduced by a factor of ρ_d since the array can be placed into a standby mode after T_{DIMA} duration.

Since the first term in both (2.20) and (2.21) is the dominant component of the energy consumption during the active mode (as C_{BL} is the largest node capacitance in either architecture), their ratio provides the energy reduction factor ρ_e as follows:

$$\rho_e = \frac{E_{DIMA}}{E_{digital}} = \frac{LB}{\beta} \tag{2.22}$$

where values for $\rho_e = 8\times$ to $32\times$ are easily achievable for typical values of $B = 4$, $L = 4, 8, 16$ and $\beta = 2$.

Hence, from (2.19) and (2.22), the EDP reduction over a digital architecture enabled by DIMA is given by:

$$\rho_{edp} = \rho_e \rho_d = \frac{(LB)^2}{\beta \gamma} \tag{2.23}$$

which ranges from $21\times$ (for $L = 4, B = 4, \beta = 2$, and $\gamma = 6$) to $1365\times$ (for $L = 16, B = 4, \beta = 1$, and $\gamma = 3$) of which the prototype IC in [52] has achieved $100\times$ EDP gains in the laboratory. This clearly indicates that there is significant room to improve upon DIMA's EDP gains achieved thus far. It is also possible to show that when comparing the energy cost of computation only, DIMA's low-swing analog computation is approximately $10\times$ lower than that of the digital architecture.

Though the energy models (2.20)–(2.21) are simple, Fig. 2.13 shows that they correlate well with measured values from silicon [32] for $N_{ROW} = 512$ and $\Delta V_{BL,max} = 500\,\text{mV}$ with a modeling error of 11%. Note that the BL capacitance $C_{BL} = C_{BLC} N_{ROW}$, where C_{BLC} is the BL capacitance per cell, i.e., C_{BL} is

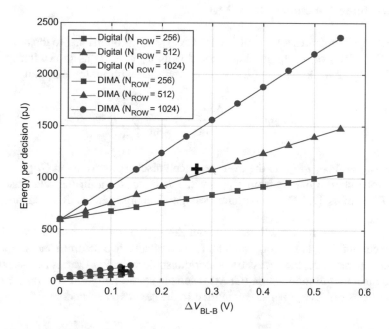

Fig. 2.13 Comparing the energy consumption of DIMA and the digital architecture obtained from (2.20) and (2.21) with $N_{ROW} = 256, 512$, and 1024, $L = 4, B = 4, \beta = 1$ in realizing a $N = 128$-dimensional SVM. The plus markers (+) indicate silicon measured results from [32] with $N_{ROW} = 512$

proportional to the number of rows N_{ROW}. Hence, Fig. 2.13 also shows that the energy consumption increases linearly with N_{ROW} and the BL swing per bit ($= \Delta V_{\text{BL-B}} = \frac{\Delta V_{\text{BL,max}}}{B}$ for DIMA and $\Delta V_{\text{BL,max}}$ for a conventional system) for both architectures because of the increased precharge energy. However, DIMA achieves enormous EDP gains by amortizing this precharge energy over the access and processing of $B \times N_{\text{COL}}$ bits compared to N_{COL}/L bits in the digital architecture. In doing so, DIMA sacrifices the SNR of its computations as discussed next.

2.5.2 Modeling the SNR of DIMA Computations

DIMA provides significant EDP gains, but these gains are obtained at the expense of the SNR of its analog computations (*compute SNR*). This section presents noise and distortion models of DIMA computations in order to relate DIMA's energy and delay to its compute SNR, and hence to the accuracy of ML algorithms realized on DIMA.

2.5.2.1 Basic Functional Read (FR)

The FR stage generates a BL voltage drop $\Delta V_{\text{BL}}(W)$ proportional to the *data value* $W = \Sigma_{i=0}^{B-1} 2^i w_i$ of a column-major stored with, $w_i \in \{0, 1\}$ as shown in Fig. 2.7a, and given by (2.8). Rewriting (2.8) with $\overline{w}_c$ replaced with w_i for simplicity,

$$\Delta V_{\text{BL}}(W) = \frac{\Delta Q_{\text{BL}}}{C_{\text{BL}}} = \frac{I_{\text{cell}}}{C_{\text{BL}}} \sum_{i=0}^{B-1} T_i W_i \qquad (2.24)$$

where $I_{\text{cell}} = V_{\text{PRE}}/R_{\text{BL}}$ is the average bitcell discharge current, $\Delta Q_{\text{BL}} = \Sigma_{i=0}^{B-1}$ is the total charge drawn from C_{BL} by the B bitcells within the total discharge time $T = \max_i \{T_i\}$. The maximum voltage discharge on the BL is denoted by $\Delta V_{\text{BL,max}} = (2^B - 1)\Delta V_{\text{BL}}(1)$.

The expression for the FR stage output $\Delta V_{\text{BL}}(W)$ in (2.24) assumes a cell discharge current I_{cell} that is both spatially (across bitcells in a column) and temporally invariant. In practice, the access transistor threshold voltage V_t varies across bitcells and channel length modulation (CLM) results in spatio-temporal variations in I_{cell}. These effects on $\Delta V_{\text{BL}}(W)$ are quantified in the next two subsections.

2.5.2.2 Temporal Variations (Distortion)

The discharge current I_{cell} varies over time because of CLM leading to deterministic but non-linear mapping from W to $\Delta V_{\text{BL}}(W)$ at the functional read output. That deterministic non-linearity is referred to as distortion and is fixed for a specific die.

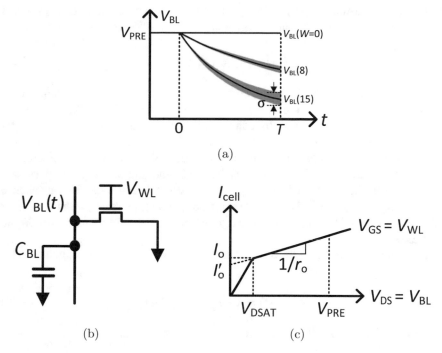

Fig. 2.14 Modeling BL discharge in the FR stage: (**a**) BL voltage $V_{BL}(t)$, (**b**) a simplified RC model, and (**c**) cell discharge current I_{cell}

Therefore, such errors can be overcome by tuning the model parameters (W) before storing them in memory, e.g., via the off-chip [53, 54] or on-chip [52] training of the inference model.

The impact of temporal variations on $\Delta V_{BL}(W)$ (Fig. 2.14a) can be evaluated by assuming a simplified circuit model of a single bitcell discharging the BL as shown in Fig. 2.14b. Assuming that the source node of the access transistor is close to 0 V during the discharge process, the cell discharge current can be approximated as (see Fig. 2.14c):

$$I_{cell}(t) = I_o + \frac{V_{BL}(t) - V_{DSAT}}{r_o} = I_o' + \frac{V_{BL}(t)}{r_o} \tag{2.25}$$

where $I_o = k_n'(V_{WL} - V_t)^\alpha$ is the current when the access transistor is at the edge-of-saturation, $1 \le \alpha \le 2$, and V_{DSAT} is the saturation drain-to-source voltage; r_o is the output resistance of the access transistor; and $I_o' = I_o - V_{DSAT}/r_o$, as shown in Fig. 2.14c.

Employing (2.25) and solving the differential equation

$$C_{BL}\frac{dV_{BL}(t)}{dt} + I_o' + \frac{V_{BL}(t)}{r_o} = 0 \tag{2.26}$$

with initial conditions $V_{BL}(0) = V_{PRE}$, we first obtain an expression for $V_{BL}(t)$, and then substitute it in the relationship $\Delta V_{BL}(t) = V_{PRE} - V_{BL}(t)$, to obtain:

$$\Delta V_{BL}(t) = (V_{PRE} + I_o' r_o)(1 - e^{-\frac{t}{\tau}}) \tag{2.27}$$

where $\tau = r_o C_{BL}$ is the RC time-constant of the discharge path. A large time-constant τ is desirable in order to reduce the impact of temporal distortion. Increasing τ by tuning r_o, e.g., by adjusting its dimensions, is limited by the tight constraints on transistor sizing in a bitcell. Though r_o can be tuned by adjusting its gate bias V_{WL}, this gate-bias tunability is constrained by other considerations such as controlling $\Delta V_{BL,max}$. On the other hand, temporal linearity improves with an increase in the number of rows N_{ROW} of the BCA because the BL capacitance $C_{BL} = C_{BLC} N_{ROW}$ (C_{BLC} is the BL capacitance per cell) increases with the number of rows in the BCA. Thus, the discharge curves in Fig. 2.14a shift up as N_{ROW} increases.

Assuming the discharge time $t \ll \tau$ and r_o is a constant, we approximate (2.27) via Taylor series expansion as:

$$\Delta V_{BL}(t) = (V_{PRE} + I_o' r_o)(t/\tau) \tag{2.28}$$

Substituting $t = T_0 \sum_i 2^i W_i$ in (2.28) results in:

$$\Delta V_{BL}(W) = \frac{(I_o + I_{CLM})T_0}{C_{BL}} \sum_{i=0}^{B-1} 2^i W_i \tag{2.29}$$

where $I_{CLM} = (V_{PRE} - V_{DSAT})/r_o$. Note that (2.29) has a form similar to that of (2.24). Therefore, we write (2.29) as:

$$\Delta V_{BL}(W) = \frac{I_{cell} T_0}{C_{BL}} \sum_{i=0}^{B-1} 2^i W_i + g(W, T) \tag{2.30}$$

where $g(W, T)$ represents the distortion that depends on the data W and total discharge time T. In practice, the distortion $g(W, T)$ is measured by calculating the percentage difference between an ideal straight line transfer function of the block and the mean of the measured values over a large number of cells.

Extracting the values for the parameters in (2.25)–(2.28) for a 65 nm CMOS process, we find that $k_n' = 220 \, \mu A/V^2$, $r_o = 74 \, k\Omega$, $I_o = 18.9 \, \mu A$, $V_t = 0.4 \, V$, $C_{BL} = 270 \, fF$, $V_{DSAT} = 0.2 \, V$, $\alpha = 1.8$, and $T_0 = 300 \, ps$. Therefore, if $V_{BL}(t)$ is allowed to range from $0.5 \, V$ to $1 \, V$, then it is straightforward to show that $I_{cell} = I_o + I_{CLM}$ in (2.24) varies up to 26 % over the total discharge time of 4.5 ns. Furthermore, approximation of (2.27) from (2.28) via a Taylor series results in a $g(W, T)$ that is at most 12 % for these values.

2.5.2.3 Spatial Variations (Noise)

In contrast, spatial variations manifest themselves as spatially distributed noise, i.e., $\Delta V_{BL}(W)$ in (2.24) can be written as:

$$\Delta V_{BL}(W) = \frac{I_{cell}}{C_{BL}} \sum_{i=0}^{B-1} T_i W_i + \mu_{BL} + \eta_{BL} \tag{2.31}$$

where μ_{BL} is the distortion (see Sect. 2.5.2.2) and η_{BL} is the spatial noise contribution. The statistics of this noise can be obtained by substituting $I_0 = k'_n (V_{WL} - V_t)^\alpha$ into (2.29) (ignoring I_{CLM} for simplicity) as follows:

$$\Delta V_{BL}(W) \propto \frac{k'_n (V_{WL} - V_t)^\alpha}{C_{BLC} N_{ROW}} \sum_{i=0}^{B-1} T_i W_i \tag{2.32}$$

where V_{WL}, the WL access voltage, is a critical variable in designing DIMA since it controls the discharge path resistance r_0 and the current. In fact, the impact of V_t variations on the discharge current I_{cell} and hence $\Delta V_{BL}(W)$ increases as V_{WL} approaches V_t. For example, the normalized standard deviation $(\sigma/\mu)_{\Delta V_{BL,max}}$ of $\Delta V_{BL,max}$ increases from 6% to 29% if V_{WL} is reduced as shown in Table 2.2. On the other hand, increasing V_{WL} can lead to read upsets (destructive reads) in and cell storing a "1" if $\Delta V_{BL,max}$ becomes excessively large, e.g., $\Delta V_{BL,max} > 0.7 V_{PRE}$.

One way to address that issue is to set T_0 to the smallest value such that the rise and fall times of the WL pulses are a small ($< 0.2 T_0$) fraction of T_0, e.g., $T_0 \approx 200$ ps to 300 ps. Once T_0 and B (the number of active rows) are fixed, then $\Delta V_{BL,max}$ can be controlled by tuning V_{WL} within the upper limit set by destructive read considerations. Note that the upper limit on V_{WL} can be increased by increasing the number of rows N_{ROW} and hence the BL capacitance (see Fig. 2.15). The reason is that the BLs discharge slower due to their higher BL capacitance, thereby allowing one to increase V_{WL} leading to reduced impact of process variations at a given $\Delta V_{BL}(W)$.

Table 2.2 Noise and distortion in DIMA stages [32]

Error type	FR (η_F)	BLP (η_B)		CBLP (η_C)
		DP	SAD	
% Distortion (μ)	2.6[a]	2.1[a]	2.5[a]	0.8[c]
% Noise (σ/μ)	(6–29)[b]	2.8[b]	3.2[b]	0.2[c]

Row 1: obtained as an average over all 16 4-b data values
Row 2: obtained for maximum discharge $\Delta V_{BL,max}$
[a]Silicon measured
[b]Monte Carlo simulations with 0.4 V $\leq V_{WL} \leq 0.8$ V
[c]Estimated from the capacitor sizes in [32]

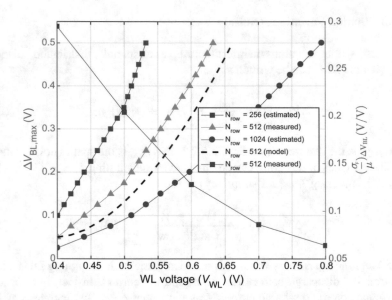

Fig. 2.15 BL swing vs. BL noise (σ_F/μ) with respect to WL voltage V_{WL}. The noise (σ_F/μ) and BL swing with $N_{ROW} = 512$ is measured from silicon prototype [32]. BL swings with $N_{ROW} = 256$ and 1024 are estimated from (2.32) and the measured BL swing with $N_{ROW} = 512$

2.5.2.4 Noise and Distortion Models

This section presents two compute-intensive distance metrics employed pervasively in ML algorithms—the dot-product and the sum-of-absolute difference (SAD)—in the presence of DIMA's non-ideal behavior described in Sects. 2.5.2.2 (distortion) and 2.5.2.3 (noise).

A digital architecture will realize (2.1) via N $B_x \times B_w$-b multiply-accumulate (MAC) operations with quantization noise as the primary source of non-ideal behavior, where B_x and B_w are the precisions of W_i and X_i, respectively.

In DIMA, because of its mixed-signal attribute, the dot product (2.1) is computed as (assuming a fixed total discharge time T):

$$\hat{y} = \frac{1}{N} \sum_{i=1}^{N} (W_i + g(W_i, T) + \eta_{wF,i}) X_i + \eta_{yB} + \eta_{yC} \qquad (2.33)$$

$$= y + \eta_y \qquad (2.34)$$

where $\eta_{F,i}$ is the spatial noise variance σ_{wF}^2 in W_i at the ith column due to FR, η_{yB} (variance σ_{yB}^2) and η_{yC} (variance σ_{yC}^2) are the noise contributions from the BLP and CBLP, respectively, as seen at the output y, and η_y is the composite of noise and distortion on y at the CBLP output, respectively (see Fig. 2.3b).

We map the FR output in (2.34) into the voltage domain in order to relate its algebraic computation to the circuit realization in DIMA, as follows:

$$\widehat{s} = \frac{1}{N}\sum_{i=1}^{N}\left(\Delta V_{\text{BL}}(W_i) + \eta_{\text{F},i}\right)x_i + \eta_{\text{B}} + \eta_{\text{C}} \tag{2.35}$$

where $\Delta V_{\text{BL}}(W_i) \in \{0, \delta, \ldots, \Delta V_{\text{BL,max}}\}$ denotes the voltage swing corresponding to the ith weight W_i with $\Delta V_{\text{BL,max}} = (2^B - 1)\delta$. The η_{F}, η_{B}, and η_{C} are the non-idealities in the voltage domain including both distortion and noise contributions from FR, BLP, and CBLP stages, respectively. $\widehat{s}$ denotes DIMA's output in the presence of the non-idealities.

Table 2.2 quantifies both distortion and noise contributions η_{F}, η_{B}, and η_{C} in the voltage domain from the FR, BLP, and CBLP stages, which indicates that the noise variances $\sigma_{\text{F}}^2 \gg \sigma_{\text{B}}^2 \gg \sigma_{\text{C}}^2$, i.e., FR noise dominates. That is to be expected since FR processing involves discharge of a BL via minimum-sized transistors in the bitcell and operates close to the near-threshold voltage regime with low V_{WL}, thereby incurring a large spatial mismatch as described in Sect. 2.5.2.3.

Similarly, the ideal SAD computation between **w** and **x**, and its noisy (DIMA) version is given by:

$$y = \frac{1}{N}\sum_{i=1}^{N}|W_i - X_i| \tag{2.36}$$

$$\widehat{y} \approx \frac{1}{N}\sum_{i=1}^{N}|W_i - X_i| + 2\eta_{\text{yF}} = y + 2\eta_{\text{yF}} \tag{2.37}$$

where the factor of 2 in (2.37) appears since both W_i and X_i are read using FR [32]. Mapping (2.37) to the voltage domain gives:

$$\widehat{s} \approx \frac{1}{N}\sum_{i=1}^{N}(|\Delta V_{\text{BL}}(W_i) - \Delta V_{\text{BL}}(X_i)| + 2\eta_{\text{F},i}) \tag{2.38}$$

$$= \frac{1}{N}\sum_{i=1}^{N}(s_i + 2\eta_{\text{F},i}) = s + \eta \tag{2.39}$$

where $|\Delta V_{\text{BL}}(W_i) - \Delta V_{\text{BL}}(X_i)| \in \{0, \delta, \ldots, \Delta V_{\text{BL,max}}\}$ denotes the voltage swing corresponding to the ith absolute difference $|W_i - X_i|$ with $\Delta V_{\text{BL,max}} = (2^B - 1)\delta$, $\widehat{s}$ denotes DIMA's output in the presence of equivalent FR noise η with variance $2\sigma_{\text{F}}^2/N$ at the CBLP output, and s is the voltage-domain version of the ideal SAD output y in (2.36).

2.5.3   Prediction of Inference Accuracy

In this section, DIMA's noise models from Sect. 2.5.2 are employed to predict its system-level classification accuracy. DIMA's accuracy is compared with that of the digital architecture operated at the same voltage swing per bit $\Delta V_{\text{BL-B}} = \frac{\Delta V_{\text{BL,max}}}{B}$. In that way, we ensure that the BL discharge energy per bit is made identical for both architectures. Two different tasks are considered: (1) template matching (TM) using the SAD kernel, and (2) SVM using the dot product kernel.

2.5.3.1   Template Matching

The TM algorithm computes the SADs between a query input $\mathbf{x}$ and a set of M candidate images $\{\mathbf{w}^{(0)}, \ldots, \mathbf{w}^{(M-1)}\}$ and outputs the index corresponding to the one with the minimum SAD, as shown below:

$$j^* = \arg \min_j y^{(j)} \tag{2.40}$$

where $y^{(j)}$ represents the SAD between $\mathbf{x}$ and the jth candidate image $\mathbf{w}^{(j)}$ as in (2.36).

- *Digital Architecture*: In the digital architecture, spatial mismatch and low $\Delta V_{\text{BL-B}}$ can result in bit flipping errors caused by insufficient input swing to the sense amplifiers. Hence, $y^{(j)}$ is computed as

$$\widehat{y}^{(j)} = \frac{1}{N} \sum_{i=1}^{N} (y_i^{(j)} + e_i) = y^{(j)} + \overline{e} \tag{2.41}$$

  where $e_i \in \{-2^B + 1, \ldots, 2^B - 1\}$ denotes the numerical error in the ith element and $\overline{e}$ denotes the sample mean of e_i:

$$\overline{e} = \frac{1}{N} \sum_{i=1}^{N} e_i \tag{2.42}$$

  where $\mathbb{E}(e_i) = 0$ and the variance $\text{Var}(e_i)$ is given by:

$$\text{Var}(e_i) = \sigma_e^2 = \left(\frac{4^B - 1}{3} \right) p \tag{2.43}$$

  where p denotes the bit error probability, which depends on $\Delta V_{\text{BL-B}}$ as follows [55]:

$$p = Q\left(\frac{\Delta V_{\text{BL-B}}}{\sigma_{\text{read}}}\right) \tag{2.44}$$

where $Q(x) = \int_x^\infty \frac{1}{\sqrt{2\pi}} \exp\left(-\frac{u^2}{2}\right) du$, and σ_{read} is a standard deviation to incorporate the effects from noise sources including the spatial variations across bitcells during the BL discharge and the SA's input offset. By the Central Limit Theorem, $\bar{e}$ can be modeled as a Gaussian for large N, i.e.,

$$\bar{e} \sim \mathcal{N}\left(0, \left(\frac{4^B - 1}{3}\right) \cdot \frac{p}{N}\right) \tag{2.45}$$

where $\text{Var}(\bar{e}) = \frac{\sigma_e^2}{N}$.

In the absence of bit flips, given two candidate images $\mathbf{w}^{(k)}$ and $\mathbf{w}^{(l)}$ and their respective SAD outputs $y^{(k)}$ and $y^{(l)}$, if $y^{(k)} < y^{(l)}$, the correct decision is $j^* = k$, since $\mathbf{w}^{(k)}$ is closer to the query image $\mathbf{x}$ than $\mathbf{w}^{(l)}$. However, in the presence of bit flips, it may turn out that $\widehat{y}^{(k)} > \widehat{y}^{(l)}$, in which case the incorrect decision $j^* = l$ will be output by the digital architecture. The mismatch probability that $\mathbf{w}^{(l)}$ is incorrectly chosen is given by (see Appendix 1):

$$p_{\text{m-digital}}^{(k\to l)} = Q\left(\alpha_{\text{TM,digital}}^{(k,l)}\sqrt{\frac{3(2^B - 1)}{2(2^B + 1)} \cdot \frac{N}{p}}\right) \tag{2.46}$$

where $0 \leq \alpha_{\text{TM, digital}}^{(k,l)} \leq 1$ denotes the normalized decision margin given by:

$$\alpha_{\text{TM,digital}}^{(k,l)} = \frac{\left|y^{(k)} - y^{(l)}\right|}{2^B - 1} \tag{2.47}$$

where $2^B - 1$ represents the maximum of the difference $|y^{(k)} - y^{(l)}|$.

Without loss of generality, suppose $j^* = 0$, i.e., $y^{(0)}$ has the minimum SAD. Then, the detection (accuracy) probability of the digital architecture is given by:

$$P_{\text{det-digital}} = \prod_{m=1}^{M-1} (1 - p_{\text{m-digital}}^{(0\to m)}) \tag{2.48}$$

where $p_{\text{m-digital}}^{(0\to m)}$ can be obtained from (2.46). We observe that $P_{\text{det-digital}}$ improves (increases) with the decision margin $\alpha_{\text{TM,digital}}^{(i,j)}$ and the vector dimension N, but worsens (reduces) if the bit error probability p increases (i.e., if $\Delta V_{\text{BL-B}}$ is reduced).

- *DIMA*: Consider two images $\mathbf{w}^{(k)}$ and $\mathbf{w}^{(l)}$ whose ideal voltage domain SAD outputs are $s^{(k)}$ and $s^{(l)}$, respectively. If $s^{(k)} < s^{(l)}$, then the correct decision is

$j^* = k$. However, in the presence of the noisy DIMA computations described by (2.39), it is possible that $\widehat{s}^{(k)} > \widehat{s}^{(l)}$, in which case an incorrect decision $j^* = l$ will be made. The mismatch probability that DIMA will incorrectly choose $\mathbf{w}^{(l)}$ (see Appendix 1) is given by:

$$p_{\text{m-DIMA}}^{(k \to l)} = Q \left(\alpha_{\text{TM,DIMA}}^{(k,l)} \sqrt{\frac{N \cdot \text{SNR}_{\text{DIMA}}}{2}} \right) \tag{2.49}$$

where the decision margin $0 \leq \alpha_{\text{TM,DIMA}}^{(k,l)} \leq 1$ in the voltage domain is given by:

$$\alpha_{\text{TM,DIMA}}^{(k,l)} = \frac{\left| s^{(k)} - s^{(l)} \right|}{\Delta V_{\text{BL,max}}} \tag{2.50}$$

which is equivalent to (2.47), and SNR_{DIMA} is defined as:

$$\text{SNR}_{\text{DIMA}} = \frac{\Delta V_{\text{BL,max}}^2}{\sigma_{\text{F}}^2} \tag{2.51}$$

As in the derivation of (2.48), the detection probability of DIMA is given by:

$$P_{\text{det-DIMA}} = \prod_{m=1}^{M-1} (1 - p_{\text{m-DIMA}}^{(0 \to m)}) \tag{2.52}$$

where $p_{\text{m-DIMA}}^{(0 \to m)}$ can be obtained from (2.49). Note that $P_{\text{det-DIMA}}$ improves with the decision margin $\alpha_{\text{TM}}^{(i,j)}$ and the dimension N, and with a sufficiently large N, accurate decisions can be made even in a low-SNR regime.

2.5.3.2 Support Vector Machine

A binary SVM computes the sign of the dot product as follows:

$$\text{sign}(s) = \text{sign}(\mathbf{w}^\mathsf{T} \mathbf{x}) \tag{2.53}$$

where the weight vector $\mathbf{w}$ is chosen to maximize the margin between the decision hyperplane and the input vectors in the training set. For ease of analysis, we assume that $\mathbf{w} = (W_1, \ldots, W_N)$ in (2.53) denotes the normalized weight vector, i.e., $0 \leq |W_i| \leq 1 \; \forall i$, and we omit the bias term.

- *Digital Architecture*: Bit flips in a digital architecture results in the transformation of the dot product computation of (2.53) into:

$$\widehat{y} = \sum_{i=1}^{N} (W_i + e_i) x_i = y + \widetilde{e} \tag{2.54}$$

where W_i is distorted to $W_i + e_i$ because of bit flips, e_i denotes the numerical error as in (2.41), and $\tilde{e}$ denotes the weighted sum of e_i, i.e.,:

$$\tilde{e} = \sum_{i=1}^{N} x_i e_i \qquad (2.55)$$

The mismatch probability of the digital architecture is given by:

$$p_{\text{m-digital}} = \Pr(\text{sign}(y) \neq \text{sign}(\hat{y})) = Q\left(\frac{N\alpha_{\text{SVM}}}{\sigma_e}\right) \qquad (2.56)$$

where $\sigma_e^2 = \left(\frac{4^B-1}{3}\right) p$ as in (2.43), and the normalized decision margin α_{SVM} is given by:

$$\alpha_{\text{SVM}} = \frac{1}{N} \cdot \left| \sum_{i=1}^{N} W_i \cdot \frac{x_i}{\|\mathbf{x}\|} \right| \qquad (2.57)$$

Note that α_{SVM} depends on the only trained weights and the input data vector (see Appendix 2 for details).

- *DIMA*: DIMA's SVM computation is well-modeled by (2.35), with signal s and noise η terms given by:

$$s = \frac{1}{N} \sum_{i=1}^{N} \Delta V_{\text{BL}}(W_i) x_i \qquad (2.58)$$

$$\eta = \frac{1}{N} \sum_{i=1}^{N} X_i \eta_{\text{F},i} \qquad (2.59)$$

where $\eta_{\text{F},i} \sim \mathcal{N}(0, \sigma_{\text{F}}^2)$. The mismatch probability of DIMA can be shown (see Appendix 2) to be:

$$p_{\text{m-DIMA}} = Q\left(N\alpha_{\text{SVM}}\sqrt{\text{SNR}_{\text{DIMA}}}\right) \qquad (2.60)$$

where $\text{SNR}_{\text{DIMA}} = \frac{\Delta V_{\text{BL,max}}^2}{\sigma_{\text{F}}^2}$.

2.5.3.3 Experimental Model Validation

In this section, the system-level accuracy prediction models (2.48), (2.52), (2.56), and (2.60), are validated by comparing their predictions of the detection probability P_{det} with silicon measured results in [32]. We consider the MIT CBCL dataset with

Table 2.3 Design and model parameters

Parameter	Values	Parameter	Values
V_{DD}	1 V	V_{WL}	0.4–0.9 V
V_{PRE}	1 V	L	4
N_{ROW}	256–1024	N_{COL}	256
T_0	300 ps	N	128–1024
B	8	M	64

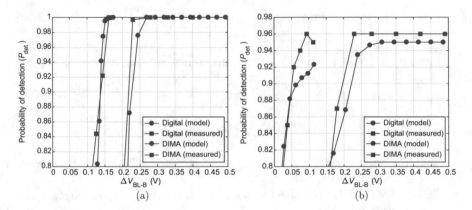

Fig. 2.16 Probability of detection (P_{det}) vs. measured results from [32] with $N_{ROW} = 512$ for: (**a**) TM, and (**b**) SVM. Note that $\Delta V_{BL,max}$ is maintained to be less than $0.7 V_{PRE}$ to avoid destructive read, e.g., $\Delta V_{BL\text{-}B} < 0.18$ V for DIMA and < 0.7 V for the digital architecture

the design parameters listed in Table 2.3, and σ/μ with respect to BL swing per bit (as shown in Fig. 2.15) to evaluate the accuracy prediction models.

For the TM algorithm, one of the 64 candidate images was chosen as the template. The accuracy P_{det} is calculated by averaging the 64 detection probabilities $P_{det\text{-}t}$ (one per template), where $P_{det\text{-}t}$ was obtained by counting the number of correct detections in multiple (>1000) trials. For SVM, 800 query images (400 faces and 400 non-faces) are tested for a face detection task and the overall P_{det} is obtained by averaging the 800 query-specific P_{det} values. Various values of the decision margin $\alpha_{TM\ or\ SVM, k \rightarrow l}$ are tried to better evaluate modeling accuracy.

Figure 2.16 shows that the predictions of system-level accuracy from ((2.48) and (2.52)) (TM) and ((2.56) and (2.60)) (SVM) match very well with values obtained from silicon [32] with a modeling error of $<10.5\%$. In general, the model estimates of accuracy are worse than the measured accuracy because these models consider the worst-case scenario.

2.5.4 Fundamental Trade-Offs and Limits

In this section, fundamental trade-offs between energy-efficiency and accuracy as a function of input vector dimension N, the decision margins (α_{TM} and α_{SVM}), the array size (N_{ROW}) for both the digital architecture and DIMA are studied. Furthermore, conditions under which DIMA provides significant EDP gains over a digital architecture, and those that limit its accuracy are identified.

The analysis in this section employ typical parameter values listed in Table 2.3. The value of $T_0 = 300$ ps was chosen to prevent a destructive read while ensuring that WL access pulses with the rise and fall times of $< 0.2T_0$ can be easily realized. The value of $\Delta V_{BL,max}$ is set by tuning V_{WL} to be less than $0.7V_{PRE}$ to avoid a destructive read, e.g., $\Delta V_{BL\text{-}B}$ <0.18 V for DIMA and <0.7 V for the digital architecture. Numerical values for each term in (2.20) and (2.21) were obtained from circuit simulations in a 65 nm CMOS process technology. The total energy estimates from (2.20) and (2.21) are validated by comparing against the measurements from the IC prototype in [32].

2.5.4.1 Energy Efficiency vs. Accuracy Trade-Offs

Figure 2.17 shows that DIMA achieves the same accuracy as the digital architecture but has an energy-per-decision cost that is lower by approximately $10\times$ for most operating conditions. Coupled with a latency reduction of $5\times$ to $20\times$ (see Sect. 2.5.1), DIMA can achieves a decision-making EDP reduction of $50\times$ to $200\times$ over a digital architecture, of which close to $100\times$ EDP reduction has already been achieved [52].

Figure 2.17 also shows that the accuracy P_{det} improves with decision margin $\alpha_{TM \text{ or } SVM}$ and input vector dimension N for the same BL swing $\Delta V_{BL\text{-}B}$. However, if the decision energy cost is kept fixed, then accuracy in fact reduces when the input dimension N is increased.

Finally, unlike that of the digital architecture, DIMA's accuracy is seen to be limited when $\alpha_{TM/SVM}$ and N are small, e.g., $\alpha_{TM} = 0.05$, $\alpha_{SVM} = 0.2$, and $N = 128$. This is because DIMA's analog computations lead to the various non-idealities introduced in Sect. 2.5.2, which have a greater impact on the accuracy of inference when $\alpha_{TM/SVM}$ and N are small.

2.5.4.2 Impact of Array Size

The number of columns N_{COL} in the bitcell array is limited only by constraints on the rise and fall times of the WL access pulses, and by the available area. However, the number of rows N_{ROW} directly impacts the system-level accuracy and energy-efficiency. This is because the BL capacitances increase in proportion to N_{ROW} requiring a higher value of the WL access pulse voltage V_{WL} to obtain the

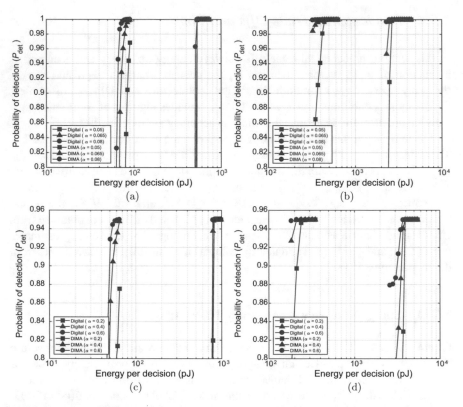

Fig. 2.17 System-level accuracy vs. energy-efficiency trade-offs obtained via accuracy prediction models ((2.48),(2.52)) by sweeping ΔV_{BL-B} for TM with $\alpha_{TM} \in [0.05, 0.08]$ and vector dimensions (**a**) $N = 128$, (**b**) $N = 512$, and via ((2.56), (2.60)) for SVM with $\alpha_{SVM} \in [0.2, 0.6]$ and vector dimensions (**c**) $N = 128$, (**d**) $N = 512$

same ΔV_{BL-B}. A higher value of the V_{WL} implies the impact of transistor threshold voltage variations on the discharge current is reduced leading to improved accuracy (see Fig. 2.18a) but at the cost of higher energy consumption (see Fig. 2.18b), for both DIMA and the digital architecture. Conversely, this also implies that N_{ROW} needs to be sufficiently large for DIMA to achieve an accuracy comparable to that of a digital architecture, e.g., Fig. 2.18 also shows that DIMA is unable to achieve maximum accuracy for $N_{ROW} = 256$. Though Fig. 2.18 shows these trends for SVM, similar trends were observed for TM as well.

2.5.4.3   DIMA Design Space

Based on the results presented in the previous sections, the following conclusions can be drawn regarding the conditions under which DIMA will perform better than a digital architecture:

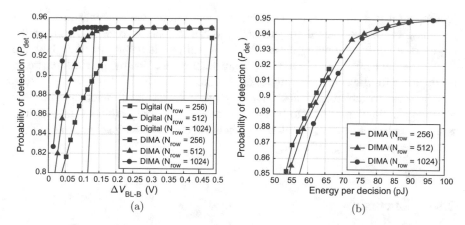

Fig. 2.18 Impact of number of rows in the bitcell array N_{ROW} on the system-level accuracy as a function of: (**a**) BL swing, and (**b**) energy per decision for SVM with $\alpha_{SVM} = 0.4$ and $N = 128$

- For a specific array size, DIMA has a minimum BL swing $\Delta V_{BL\text{-}B}$ to achieve maximum accuracy, e.g., Fig. 2.16 shows this to be $\approx 100\,mV$ for DIMA and $\approx 250\,mV$ for the digital architecture for $N_{ROW} = 512$.
- DIMA's decision accuracy improves with higher values of the decision margin $\alpha_{TM/SVM}$ at the same decision energy due to the intrinsic robustness of the ML algorithm, and for all values of the decision margin, DIMA consumes approximately $10\times$ less energy than a digital architecture for the same decision accuracy (see Fig. 2.17).
- DIMA's decision accuracy can be improved by increasing N at the same $\Delta V_{BL\text{-}B}$ and decision margin, but at a higher energy cost, e.g., approximately $4\times$ more energy when increasing N from 128 to 512, which is still much less than that of the digital architecture at the same accuracy.
- DIMA is unable to achieve an accuracy comparable to that of a digital architecture when $\alpha_{TM\ or\ SVM}$, N and/or N_{ROW} are small, e.g., $\alpha_{TM} = 0.05$ and $N = 128$, or when both N_{ROW} and N are small, e.g., $N_{ROW} = 256$ and $N = 128$.

Therefore, we can conclude that DIMA is favorable when the vector length N and the number of rows N_{ROW} are large enough for a classification task with moderate difficulty, i.e., moderate values of $\alpha_{TM\ or\ SVM}$. Since ML algorithms tend to have high inherent error immunity (i.e., large $\alpha_{TM\ or\ SVM}$) and require a number of model parameters to be stored, one can expect that DIMA will continue to exhibit high decision-level EDP gains over the digital architecture in most scenarios.

2.5.4.4 Impact of Technology Scaling

In advanced CMOS process nodes, we expect to see improved energy and delay due to reduced capacitance and increased I_{cell}. However, advanced nodes also exhibit increased process variations which will be reflected in a higher σ/μ of $\Delta V_{BL,max}$

leading to a loss in accuracy. From (2.51) and (2.60), we find that this loss in accuracy can be recovered by increasing either N and/or $\Delta V_{\text{BL,max}}$, but at the expense of increased energy costs. Hence, it will be interesting to study how the decision-level EDP gains due to technology scaling are offset by the mechanisms to compensate for the corresponding loss in accuracy.

The accuracy model indicates that larger vector length and decision margin allow analog operations in DIMA to be more accurate and energy-efficient. In addition, a large number of rows in bitcell array leads to better system-level accuracy at the cost of degraded energy efficiency.

2.6   Conclusion

This chapter describes DIMA's four processing stages: (1) FR, (2) BLP, (3) CBLP, and (4) thresholding (RDL), based on the common functional flow of ML algorithms. DIMA's accuracy in the low-SNR regime was demonstrated with design principles and measurement results for a prototype IC. The design guidelines and techniques provided in this chapter has been applied to two prototype ICs, which will be introduced in the following chapters. This chapter also analyzed DIMA in terms of its decision energy, delay, and accuracy via theoretical modeling and analysis, validation with measured results of the published silicon IC prototypes [32, 52], and identified conditions under which DIMA will be superior to a digital architecture.

The DIMA energy, delay, and accuracy models presented in this chapter, can be employed to study the benefits of a DIMA-based system for other bitcell architectures (8T or 10T) in the context of new ML algorithms and applications, e.g., natural language processing using long-short term memory (LSTM) and/or process technologies (e.g., CMOS at smaller nodes, resistive RAM and magnetic RAM). Furthermore, given DIMA's regular structure, these models, along with design principles, can be encapsulated into platform design tools such as a DIMA memory compiler to automatically synthesize DIMA macros.

Appendix 1: Template Matching

Conventional Digital Architecture

Suppose that there are two images $\mathbf{w}^{(k)}$ and $\mathbf{w}^{(l)}$ such that $y^{(k)} < y^{(l)}$. The mismatch probability that $\mathbf{w}^{(l)}$ is incorrectly chosen instead of $\mathbf{w}^{(k)}$ is given by:

$$p_{\text{m-digital}}^{(k \to l)} = P(\widehat{y}^{(l)} > \widehat{y}^{(k)})$$

$$= P(y^{(l)} - y^{(k)} < \overline{e}^{(k)} - \overline{e}^{(l)}) \tag{2.61}$$

where $y^{(l)} - y^{(k)}$ denotes the decision margin and $\overline{e}^{(k)} - \overline{e}^{(l)}$ denotes the effective error in TM's digital computation. If the effective error is greater than the decision margin, then a mismatch occurs.

If $\overline{e}^{(k)}$ and $\overline{e}^{(l)}$ are independent, then $\left(\overline{e}^{(k)} - \overline{e}^{(l)}\right)$ can be modeled by Gaussian distribution follows:

$$\left(\overline{e}^{(k)} - \overline{e}^{(l)}\right) \sim \mathcal{N}\left(0, 2 \cdot \left(\frac{4^B - 1}{3}\right) \cdot \frac{p}{N}\right) \tag{2.62}$$

and the mismatch probability is given by:

$$p_{\text{m-digital}}^{(k \to l)} = Q\left(\frac{\left|y^{(k)} - y^{(l)}\right|}{\sqrt{\text{Var}(\overline{e}^{(k)} - \overline{e}^{(l)})}}\right)$$

$$= Q\left(\sqrt{\frac{N}{p}\left(\frac{3}{2(4^B - 1)}\right)} \cdot \left|y^{(k)} - y^{(l)}\right|\right) \tag{2.63}$$

Substituting (2.47) into (2.63) results in the final expression for (2.46) the mismatch probability:

$$p_{\text{m-digital}}^{(k \to l)} = Q\left(\alpha_{\text{TM,digital}}^{(k,l)} \sqrt{\frac{3(2^B - 1)}{2(2^B + 1)} \cdot \frac{N}{p}}\right) \tag{2.64}$$

DIMA

Suppose that there are two images $\mathbf{w}^{(k)}$ and $\mathbf{w}^{(l)}$ such that $s^{(k)} < s^{(l)}$. Similar to (2.63), we can derive the mismatch probability that $\mathbf{w}^{(l)}$ is incorrectly chosen instead of $\mathbf{w}^{(k)}$ by DIMA's TM computation as follows:

$$p_{\text{m-DIMA}}^{(k \to l)} = P(\widehat{s}^{(k)} > \widehat{s}^{(l)}) = P(s^{(l)} - s^{(k)} < \overline{n}^{(k)} - \overline{n}^{(l)})$$

$$= Q\left(\sqrt{\frac{N}{2\sigma_F^2}} \cdot \left|s^{(k)} - s^{(l)}\right|\right) \tag{2.65}$$

which comes from $\left(\overline{n}^{(k)} - \overline{n}^{(l)}\right) \sim \mathcal{N}\left(0, \frac{2\sigma_F^2}{N}\right)$. Substituting (2.50) into (2.65) results in final expression (2.49) for the mismatch probability, shown below:

$$p_{\text{DIMA}}^{(k \to l)} = Q \left(\alpha_{\text{TM,DIMA}}^{(k,l)} \sqrt{\frac{N \Delta V_{\text{BL,max}}^2}{2\sigma_F^2}} \right)$$

$$= Q \left(\alpha_{\text{TM,DIMA}}^{(k,l)} \sqrt{\frac{N \cdot \text{SNR}_{\text{DIMA}}}{2}} \right) \tag{2.66}$$

where $\text{SNR}_{\text{DIMA}} = \frac{\Delta V_{\text{BL,max}}^2}{\sigma_F^2}$.

Appendix 2: Support Vector Machine

Conventional Digital Architecture

The mismatch probability for an SVM is given by:

$$p_{\text{m-digital}} = \text{Pr}(\text{sign}(y) \neq \text{sign}(\widehat{y})) = \text{Pr}(|y| < \widetilde{e}) \tag{2.67}$$

Note that $\mathbb{E}(\widetilde{e}) = 0$ because of $\mathbb{E}(e_i) = 0$. The variance of $\widetilde{e}$ is given by:

$$\text{Var}(\widetilde{e}) = \sum_{i=1}^{N} x_i^2 \cdot \sigma_e^2 = \|\mathbf{x}\|_2^2 \cdot \sigma_e^2 \tag{2.68}$$

where $\sigma_e^2 = \left(\frac{4^B - 1}{3} \right) p$ as in (2.43).

By the Central Limit Theorem, we claim that

$$\widetilde{e} \sim \mathcal{N} \left(0, \|\mathbf{x}\|_2^2 \cdot \sigma_e^2 \right) \tag{2.69}$$

Then, the mismatch probability is given by:

$$p_{\text{m-digital}} = Q \left(\frac{|y|}{\sqrt{\text{Var}(\widetilde{e})}} \right) = Q \left(\frac{|y|}{\|\mathbf{x}\|_2 \sigma_e} \right)$$

$$= Q \left(\frac{\left| \sum_{i=1}^{N} W_i \cdot \frac{x_i}{\|\mathbf{x}\|_2} \right|}{\sigma_e} \right)$$

$$= Q \left(\frac{N \cdot \alpha_{\text{SVM}}}{\sigma_e} \right) \tag{2.70}$$

where the normalized decision margin is given by $\alpha_{\text{SVM}} = \frac{1}{N} \cdot \left| \sum_{i=1}^{N} W_i \cdot \frac{x_i}{\|\mathbf{x}\|} \right|$, which is the same as (2.56) and (2.57).

DIMA

Since $n_{\text{F},i} \sim \mathcal{N}(0, \sigma_{\text{F}}^2)$, we obtain $\mathbb{E}(\eta) = 0$, and the variance of η is given by:

$$\text{Var}(\eta) = \frac{1}{N^2} \sum_{i=1}^{N} |X_i|^2 \sigma_{\text{F}}^2 = \frac{\|\mathbf{x}\|_2^2 \sigma_{\text{F}}^2}{N^2} \tag{2.71}$$

Hence, $\eta \sim \mathcal{N}\left(0, \frac{\|\mathbf{x}\|_2^2 \sigma_{\text{F}}^2}{N^2}\right)$ and the mismatch probability of DIMA in (2.60) can be derived as:

$$
\begin{aligned}
p_{\text{m-DIMA}} &= Q\left(\frac{|s|}{\sqrt{\text{Var}(\eta)}}\right) \\
&= Q\left(\frac{\left|\sum_{i=1}^{N} \Delta V_{\text{BL}}(W_i) \cdot \frac{x_i}{\|\mathbf{x}\|}\right|}{\sigma_{\text{F}}}\right) \\
&= Q\left(\frac{\left|\sum_{i=1}^{N} W_i \cdot \frac{x_i}{\|\mathbf{x}\|_2}\right| \cdot \Delta V_{\text{BL,max}}}{\sigma_{\text{wF}}}\right) \\
&= Q\left(N\alpha_{\text{SVM}} \cdot \sqrt{\text{SNR}_{\text{DIMA}}}\right) \tag{2.72}
\end{aligned}
$$

where $\Delta V_{\text{BL}}(W_i) = W_i \Delta V_{\text{BL,max}}$ due to the normalized trained weights W_i.

Chapter 3
DIMA Prototype Integrated Circuits

Chapter 2 demonstrated DIMA's versatility by enabling various vector distance (VD) computations and demonstrating significant energy and delay benefits in simulations. This chapter presents the design of two DIMA prototype ICs: (1) a multi-functional DIMA [44], and (2) a random forest (RF) DIMA [42] in a 65 nm process. The multi-functional DIMA IC supports four different algorithms: the support vector machine (SVM), template matching (TM), k-nearest neighbor (k-NN), and matched filter (MF). The RF DIMA IC employs an ensemble of decision trees for classification. For both prototypes, design details, including the chip architecture, circuit techniques, and measured results, are provided.

3.1 The Multi-Functional DIMA IC

The multi-functional DIMA IC realizes four algorithms based on computation an appropriate VD between the stored data ($\mathbf{w}$) and the input data ($\mathbf{x}$). The prototype IC (Fig. 3.1) is designed in a 65 nm CMOS process and packaged in an 80-pin quad flat no-leads package (QFN).

3.1.1 Architecture

The chip architecture in Fig. 3.2 comprises a DIMA core (CORE), a digital controller (CTRL), and an input register to stream in the operand $\mathbf{x}$. The CORE includes a 512×256 BCA, the conventional SRAM read/write circuitry, the BLP and CBLP, and four 8-b single-slope ADCs [56]. The RDL is embedded in the CTRL. The SRAM bitcell was custom-designed following standard layout design rules as the industry memory compiler did not permit modifications to the peripheral

© Springer Nature Switzerland AG 2020
M. Kang et al., *Deep In-memory Architectures for Machine Learning*,
https://doi.org/10.1007/978-3-030-35971-3_3

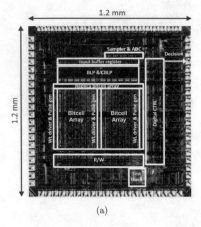

Technology	65 nm CMOS
Die size	1.2 mm × 1.2 mm
CTRL operating freq.	1 GHz
SRAM capacity	16 KB (512 × 256)
Bitcell dimension	2.11 × 0.92 um^2
Supply voltage	CORE: 1.0 V, CTRL: 0.85 V

(a)   (b)

Fig. 3.1 The multi-functional DIMA IC: (**a**) die micrograph, and (**b**) chip summary

circuitry. As a result, the horizontal and vertical dimensions of the bitcell were approximately $1.7\times$ larger than than those of a typical foundry-provided bitcells [57]. A column-muxing ratio with $L = 4$ was chosen to maximize the throughput for the standard SRAM read.

An 8-b precision is chosen for W and X in order to maintain an accuracy close to (<1% loss) the digital fixed-point accuracy [1, 58, 59]. Based on the design principles in Sect. 2.4.1.1, parameter values $V_{\mathrm{WL}} = 0.65$ V and $T_0 \approx 250$ ps were chosen resulting in the longest PWM-WL pulse width $T_3 < 0.4 R_{\mathrm{BL}} C_{\mathrm{BL}}$ to ensure sufficient linearity and avoid read upsets. A serially provided reconfiguration word (*RCFG*) initializes the local controllers in CTRL.

The circuitry for normal read and write operations (R/W block in Fig. 3.1a) occupies 14% of the CORE, which includes the SAs, write drivers, and the column decoder. On the other hand, the total area overhead due to DIMA circuitry was found to be 19% of the CORE area. Specifically, the charge-based multiplier (Fig. 3.3a) for the DP mode comprises 10%, and the analog comparator, mux, and replica BCA (Fig. 3.3b) for the MD mode takes up 9%. The CBLP does not require additional area, as it shares its circuitry with BLP blocks. Support for the FR functionality does not incur an additional area penalty, as it employs pre-existing WL drivers.

3.1.2 Timing

The chip operations are sequenced via the CTRL which operates with a master 1 GHz clock (CLK) thereby providing a 1 ns time resolution to generate control signals for the CORE. Self-timed control [60, 61] can improve the throughput of both normal mode and DIMA mode operations but a synchronous design was chosen for simplicity.

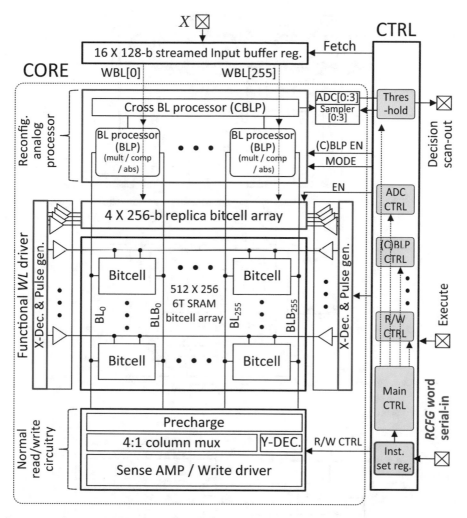

Fig. 3.2 The multi-functional DIMA IC architecture

The timing diagram in Fig. 3.4a describes a sequence of ten events that occur during a *word-row period*, i.e., when processing a single word-row of $B = 8$ bits through the FR, BLP, and CBLP stages. The first event in both the MD and DP modes is the BL precharge. Next, the FR, BLP, and CBLP stages are sequentially executed to generate the corresponding outputs V_{BL}, V_B, and V_C, respectively. One difference between the two modes is that only the MD mode requires transfer of **x** from the input buffer into replica BCA before FR is initiated. On the other hand, the DP mode needs to make that transfer of **x** to the mixed-signal multiplier before initiating BLP and requires additional delay in the CBLP stage to support sub-ranged processing. The last event samples the CBLP output to generate the input

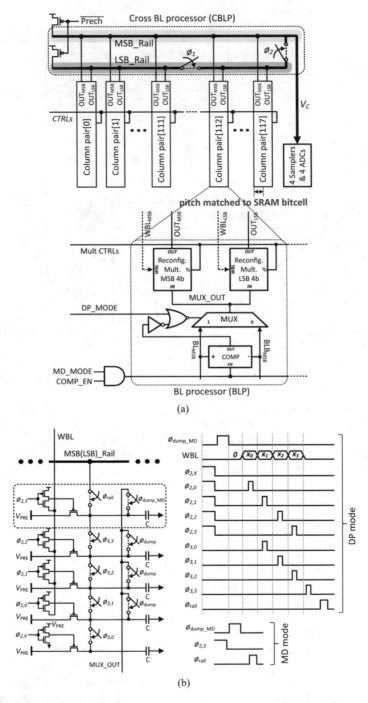

Fig. 3.3 BLP and CBLP implementations for reconfiguration: (**a**) overall structure, and (**b**) reconfigurable charge-based multiplier for 4-b MSB (or LSB) and its enable signals (MD mode uses only the circuits in the dotted RED box)

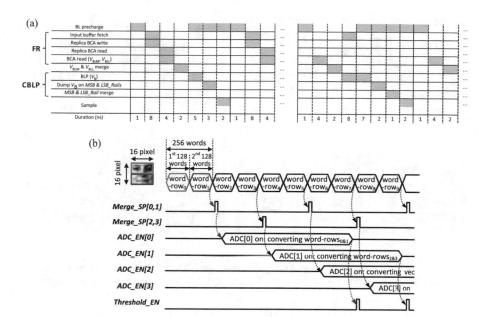

Fig. 3.4 DIMA timing for: (**a**) a single word-row, and (**b**) multiple word-rows (dotted RED line shows a single thread to process 256 words)

voltage for the ADC. Each event requires an integer number of CLK cycles, which are estimated via post-layout simulations. The prototype IC provides tunability to allow additional CLK cycles for each stage in order to accommodate deviations from the nominal process corner.

Figure 3.4b shows the timing diagram for processing 256-dimensional **w** and **x**. Each word-row consists of 128 8-b words, since a sub-ranged read results in a 128-dimensional vector even though $N_{COL} = 256$. The result of FR is the generation of the CBLP output V_C. Two word-rows are processed consecutively and their outputs V_Cs are sampled and charge-shared (*Merge_SP* step) to aggregate 256 SD results. Then 1-of-4 ADCs are used to digitize the analog V_C into an 8-b word. The single-slope ADC conversion takes 140 CLK cycles for both the MD and DP modes, which is approximately 5.6 MD word-row periods. However, that slow conversion rate is not an issue, as four ADCs operate in parallel. The ADC output is further processed in the RDL block to realize the thresholding operation (*Threshold_EN* step).

3.1.3   Algorithm and Application Mapping

Four tasks: (1) face detection using SVM; (2) gun shot detection using MF; (3) face recognition using TM; and (4) handwritten digit recognition using k-NN (see Fig. 3.5) were mapped on to the prototype IC. Those tasks cover both binary and

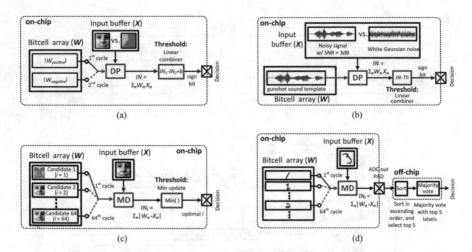

Fig. 3.5 Four inference tasks mapped on to the prototype IC: (**a**) SVM for face detection, (**b**) MF for event detection, (**c**) TM for face recognition, and (**d**) k-NN for handwritten number recognition

multi-class (4-class and 64-class) scenarios, requiring both MD and DP modes of operation, and processing of both image and sound datasets [62–64] as summarized in Table 3.1. Table 3.2 defines the set of operations in each stage that can be chosen using the *RCFG* word. In that process, the prototype IC is able to realize the four different algorithms as shown in Table 3.3.

In MF (Fig. 3.5b), the final decision is generated right after a single DP processing step and thresholding. On the other hand, SVM requires signed coefficients. Thus, the absolute values of the positive and negative coefficients are stored in separate rows (Fig. 3.5a). and processed in consecutive cycles. These are then compared in the RDL stage to obtain the sign of the DP. TM (Fig. 3.5c) and k-NN (Fig. 3.5d) make decisions after comparison (to find the minimum) or majority voting across multiple candidates. All the datasets are processed fully on-chip except for k-NN, for which the final step of majority voting is done off-chip.

3.2   Measured Results

This section describes measured results from the prototype IC in terms of its energy, delay, and accuracy, both at the local stage level and at the inference task level.

Table 3.1 Datasets used for chip measurements [62–64]

	Task	# of classes	Algorithm	Dataset	Remarks (X: query input, W: data stored in array)
1	Face detection	2	SVM	MIT CBCL dataset	– 100 query inputs tested – W: feature extractor and classifier combined 23×22 8-b coefficient – X: 23×22 8-b pixel image (face/non-face)
2	Event (gun shot) detection	2	MF	Gun shot Sound	– 100 query inputs tested – W: gun shot mono sound data with 256 8-b words – $X1$: gun shot sound contaminated by AWGN with 3 dB SNR or $X2$: Only AWGN with equal power of "signal + AWGN" in $X1$
3	Face recognition	64	TM	MIT CBCL dataset	– 64 query inputs tested (due to array size limit) – 16×16 8-b pixel image for X and W – W: 64 candidate faces, X: one of the 64 candidate faces in W
4	Handwritten number recognition	4	k-NN	MNIST dataset	– 100 query inputs tested – 4 classes from "0" to "3" (due to array size limit) – 16×16 8-b pixel image for X and W, W: 16 images per class, X: image from 4 classes

3.2.1 Accuracy of FR

The measured results for the sub-ranged FR of 8-b the word W is shown in Fig. 3.6a. The ΔV_{BL} generated by FR for all 256 values of W was measured at the output of the column mux along the normal SRAM read path. The sudden jump when W transitioned from 7 to 8 was due to the large change in the average portion of transition time in the WL pulses. The overall integral non-linearity (INL) was found to be less than 0.87 LSB.

Table 3.2 Choices of functions at each processing stage

Stage	Configurations
FR	① Normal read ② Digital to analog conversion ③ Scalar ADD or SUBT
BLP	① Scalar MULT ② *BL*-wise sampling ③ Absolute value
CBLP	① Aggregation ② Weighted aggregation
RDL	① MIN or MAX ② Linear combination ③ Send outside chip

Table 3.3 Configurations per stage to enable four algorithms. The functions corresponding to numbers are described in Table 3.2

Mode	Algorithm	FR	BLP	CBLP	RDL
DP	SVM	②	①, ②	②	②
	MF	②	①, ②	②	②
MD	*k*-NN	②, ③	②, ③	①	③
	TM	②, ③	②, ③	①	①

The variation in ΔV_{BL} was measured (see Fig. 3.6b) by using the stored data $W_M = W_L = 7$ is chosen to generate the worst case (maximum) variation in ΔV_{BL}, as in this case, the BL is discharged by a single SRAM bitcell without the benefit of averaging. Figure 3.6b shows that the variation in ΔV_{BL} has a standard deviation $\sigma = 2.5\,\text{mV}$ ($\sigma/\mu = 1.1\%$), which is only 0.6% of the dynamic range (410 mV) of CBLP output V_C in this test mode. That small deviation arose because of the averaging effect during the aggregation of the 128 BLP outputs in the CBLP stage. In the following sections, we show that such errors have a negligible impact on the inference accuracy.

3.2.2 Accuracy of CORE Output

We characterize the accuracy of the CORE output that includes FR, BLP and CBLP stages as shown in Fig. 3.7. The measured error magnitudes at V_C (from an ideal

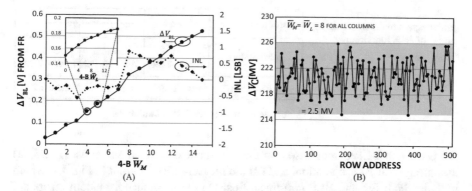

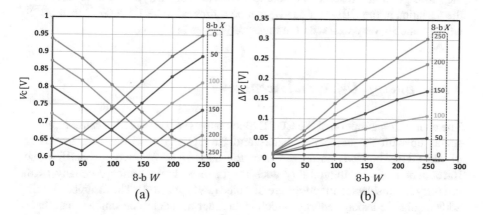

Fig. 3.6 Measured FR accuracy of 8-b W: (**a**) BL voltage drop ΔV_{BL} with sub-ranged read, and (**b**) the impact of spatial variations on the CBLP output voltage drop ΔV_C, which measured by first storing the same data ($W_M = W_L = 7$) across the entire BCA, accessing the word-rows via FR, and then aggregating via the CBLP

Fig. 3.7 Measured CORE analog output with 8-b operands W and X in the: (**a**) MD mode ($\sum |W_i - X_i| \propto V_C$), and (**b**) the DP mode ($\sum W_i X_i \propto \Delta V_C$), where the same data W and X are stored in all the columns

linear trend) in the DP and MD modes are <18 mV and <28 mV with a mean of 4 mV and 8 mV, respectively, over all (W, X) combinations. Though these errors are significantly larger than the chosen target resolution of $V_{res} = 1$ mV, these errors do not affect the accuracy of inference tasks which have sufficient decision margins (as discussed next). In addition, it is possible to obtain circuit-aware model hyperparameters by training in the presence of these errors.

Table 3.4 summarizes the accuracy of each function per stage in terms of deterministic and random error contributions. Deterministic error contributions arise from the inherent non-linearity of the proposed circuits, whereas random error contributions arise from process variations. As it is difficult to isolate the BLP stage from the FR stage, the deterministic error of BLP is obtained by comparing (1)

Table 3.4 Error magnitude (with respect to the dynamic range) of each function per stage

Error magnitude	FR	BLP	
		DP	MD
Deterministic	max: 5.8% (mean: 2.6%)	max: 6% (mean: 2.1%)	max: 7.5% (mean: 2.5%)
Random $(\sigma/\mu)^a$	max: 12.9%	max: 2.8%	max: 3.2%

[a] Obtained from Monte Carlo simulations

the estimated BLP output based on the measured non-linear FR curve (Fig. 3.6a) assuming ideal BLP operation, and (2) the measured BLP output in Fig. 3.7, which includes both FR and BLP non-linearities. On the other hand, random errors are obtained via Monte Carlo simulations, as it is difficult to extract the variations across a statistically significant number of BLP units. Each function generates a deterministic error of less than 3% of the output dynamic range on average, whereas the FR stage creates a dominant random error exhibiting a $\sigma/\mu = 12.9\%$. Therefore, the variation at the CBLP output V_C can be estimated to be 1.1% ($=12.9\%/\sqrt{128}$), which matches very well with the measured values in Fig. 3.6b.

3.2.3 Energy, Delay, and Accuracy

The energy consumption of the CORE block only was considered because its energy scales up with the number of banks and the BCA size. In contrast, the energy of the CTRL block is amortized over the number of banks and the BCA size. Introduction of reconfigurability does not incur additional energy penalty as the circuitry for unselected functions are disabled (and bypassed). The disabled circuitry adds negligible leakage energy, which is an order of magnitude smaller than that of the BCA.

The CORE decision energy and decision accuracy was measured for SVM (face detection, binary class) and TM (face recognition, 64-class) tasks. The CORE decision energy was normalized by the number of 8-b data words processed per decision to obtain the energy-per-word as a function of BL swing per LSB $\Delta V_{lsb} = \Delta V_{BL}(\overline{X}_M = 15)/15$ in Fig. 3.8a. Figure 3.8a indicates that CORE energy reduces at a rate of 0.2 pJ (0.4 pJ) per 20 mV for a binary or DP mode (64-class or MD mode) task. The higher slope of the energy vs. ΔV_{lsb} curve in the MD mode and its higher energy consumption for $\Delta V_{lsb} > 15$ mV result from the MD mode's use of the replica BCA, which causes an additional voltage drop on the BL during FR.

The accuracy of the inference task is measured by the probability of detection (P_{det}) obtained by normalizing the number of correctly classified queries by the total number of queries. Figure 3.8b shows that the binary task is more robust than the 64-class task at the same ΔV_{lsb}. Furthermore, the binary and 64-class tasks achieve >90% detection accuracy for $\Delta V_{lsb} > 15$ mV and $\Delta V_{lsb} > 25$ mV, respectively.

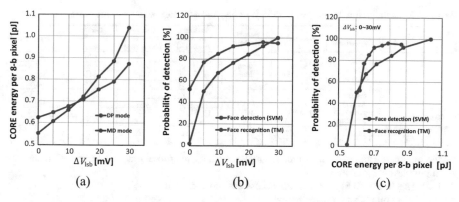

(a) (b) (c)

Fig. 3.8 Measured trade-offs between: (**a**) CORE energy vs. BL swing per LSB (ΔV_{lsb}), (**b**) decision accuracy (P_{det}) vs. ΔV_{lsb}, and (**c**) P_{det} vs. CORE energy

Fig. 3.9 CORE energy consumption of DIMA ([†] post-layout simulations, [‡] measured) compared to a conventional (Conv.) digital architecture (*SRAM energy measured and digital computation energy from post-layout simulations)

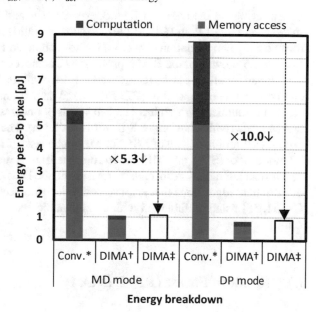

Figure 3.8c plots the detection accuracy against the CORE energy per 8-b pixel and shows the accuracy and energy trade-off.

Next, the DIMA prototype is compared with a conventional 8-b digital reference architecture (REF). REF is a 2-stage pipelined architecture comprising an SRAM of the same size as the one in the DIMA prototype, and a digital block synthesized separately for realizing an SVM (DP mode) and a TM (MD mode). The energy and delay of the digital block in REF were estimated from post-layout simulations. The energy and delay of the SRAM in REF were measured from the DIMA prototype in the normal read mode. Figure 3.9 shows the energy breakdown for REF, DIMA (post layout simulations of the DIMA), and the DIMA prototype IC. The measured

Table 3.5 Comparison of energy efficiency, delay, and accuracy with prior art

	Process (nm)	# of algorithms	Memory size	Input bit precision	Decision throughput (Decisions/s)	Decision energy (pJ/decision)	Decision EDP (f J·s)	Accuracy
This work	65 CMOS	4 (SVM, MF, k-NN, TM)	SRAM 512 X 256-b	W: 8b X: 8b	SVM: 9.3M	446	0.05	95 %
					MF: 18.5M	223	0.01	100 %
					TM: 312.5K	16.9K	54.0	100 %
					KNN: 312.5K	16.9K	54.0	92 %
8-b digital[a] (REF)	65 CMOS	synthesized dedicated processor per algorithm	SRAM 512 X 256-b	W: 8b X: 8b	SVM: 1.7M	4.5K	2.6	96 %
					MF: 3.4M	2.2K	0.6	100 %
					TM: 54.3K	93.0K	1715.3	100 %
					KNN: 54.3K	93.0K	1715.3	90 %
[58][b]	14 Tri-gate	1 (k-NN)	128 byte	W: 8b, X: 8b	21.5M	3.4K	0.2	Not reported
[31][c]	130 CMOS	1 (Adaboost)	SRAM 128 X 128-b	W: 1b, X: 5b	50M	633.4	0.01	90 %

[a] Memory energy and delay measured from prototype IC, but those of digital computation obtained from post-layout simulations
[b] Single function with SRAM memory access cost not included
[c] Single function with 1b weight vector

energy savings in the DP and MD modes were 10× and 5.3×, respectively, due to small swing FR, BLP, and CBLP. Furthermore, the DIMA energy estimates obtained from post layout simulations are close to those obtained from measurements.

Table 3.5 shows that the DIMA prototype IC achieves negligible (≤1%) accuracy degradation for all four tasks relative to REF. In addition, DIMA requires 16× fewer read accesses as compared to REF for a fixed data volume. The reason is DIMA's massive parallelism (which has 128 8-b words per access) compared to the normal SRAM mode (which has only 8 8-b words per fetch). Thus, the DIMA prototype IC provides a throughput gain of 5.8× for MD mode tasks (TM and k-NN) and 5.3× for DP mode tasks (SVM and MF). Therefore, the EDP is reduced by 32× and 53× in the MD and DP modes, respectively. As a result, the DIMA prototype IC implements four different algorithms that achieve better decision accuracy than single function ICs [31, 58] listed in Table 3.5 with a comparable energy-delay product (scaled to 65 nm).

3.3   Random Forest (RF) DIMA IC

This section presents an IC realization of a random forest (RF) ML classifier that is based on the DIMA platform [42]. The RF classifier [65] is attractive due to its high-accuracy, simple operations (i.e., comparisons), applicability to multi-class problems, and robustness to non-ideal computations due to its majority-voting based decision making. However, realization of an energy-efficient implementation of the RF algorithm is challenging due to its high data access rate combined with its highly irregular data access pattern. This section presents an energy-efficient and high throughput RF classifier IC that employs (1) deterministic subsampling (*DSS*) to reduce interconnect complexity, (2) a balanced decision tree to regularize memory access pattern, and (3) deeply embedded analog computations [44] in the periphery of an SRAM bitcell array (BCA) to exploit the inherent algorithmic error tolerance. To the best of our knowledge, this is the first IC implementation of the RF algorithm as earlier works are limited to FPGAs, GPUs, and multi-core processor implementations of the RF algorithm [65]. These works fail to take advantage of the opportunities afforded by analog computations.

3.3.1   The Random Forest (RF) Algorithm

The RF algorithm (Fig. 3.10) consists of M decision trees, each comprising a maximum of N nodes where $1 + \log_2(N + 1)$ is the tree depth, including the leaf nodes. The mth tree processes data obtained by random sub-sampling (RSS) of the input image $\mathbf{x}$ based on a pseudo-random pattern vector $\mathbf{p}_m$, which is obtained during the training stage. The nth node in the mth tree compares the pixel (or feature) $x(p_{m,n})$ indexed by $p_{m,n}$ with a threshold $\tau_{m,n}$ to obtain a node-level binary decision $q_{m,n}$. Either the left or right branch is taken based on $q_{m,n}$. This process is repeated until a leaf node is reached. The label $c_{m,l}$ corresponding to the lth leaf node is the tree-level decision. The final decision is obtained by majority-voting M such tree-level decisions. Although the RF algorithm is conceptually simple, it has a number of implementation challenges, as described next.

3.3.2   RF Implementation Challenges

To enable an energy-efficient and low-latency RF system, the following implementation challenges need to be addressed:

1. *Complex crossbar*: The RSS operation requires a complex crossbar (e.g., $K{:}1$, where $K > 256$ is the number of pixels in $\mathbf{x}$) to route a specific pixel of $\mathbf{x}$ to the corresponding tree node, i.e., to generate $x(p_{m,n})$ in Fig. 3.10.
2. *Irregular trees*: Each tree can have a different shape and different number of nodes (e.g., up to $2^{(N-1)}$ possible shapes with $N \geq 31$) as shown in Fig. 3.10. The $q_{m,n}$s from each tree node are employed as the address to a look-up-table (LUT) to obtain $c_{m,l}$. The irregular tree shapes incur different processing delays and complex LUT logic (e.g, requires don't care logic for non-existent tree nodes).
3. *Parallelizing comparisons*: Low-latency requires computations of many tree nodes in parallel including the nodes on the non-selected paths. This requires highly parallel comparisons by fetching $2N$ bytes data and N comparisons per tree (e.g., $N \geq 31$) via the limited bit-widths of SRAM IO (B_{IO}) and the bus (B_{BUS}) (e.g., $B_{IO} = B_{BUS} = 64$-b per SRAM bank) for 64 to a few hundred trees.
4. *LUT inefficiency*: A brute-force implementation of the LUT to compute the labels $c_{m,l}$ from tree node decisions $q_{m,n}$ requires fetching to the LUT contents of 2^N

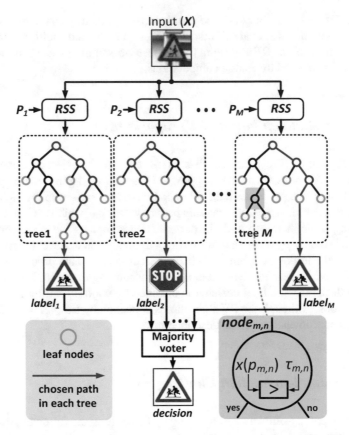

Fig. 3.10 The random forest (RF) algorithm showing the random sub-sampling (RSS), decision tress, and majority voting stages

cases (e.g., $N \geq 31$), each consisting of an index and corresponding label. The LUT computations need to be processed in additional hardware.

The above listed challenges are addressed as follows. (a) Challenges 1 and 2 are addressed by a modified RF algorithm (in Fig. 3.11) that uses regularized trees and deterministic sub-sampling (DSS), respectively; (b) DIMA is used for implementation of tree node comparisons (Challenge 3); (c) A class ADD generator (CAG) is used to address Challenge 4. As a result, the required number of computations is significantly reduced as seen in Table 3.6.

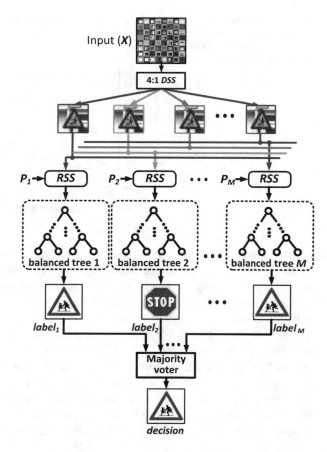

Fig. 3.11 The proposed modifications to the RF algorithm with deterministic sub-sampling (DSS), and regularized decision trees

Table 3.6 Required operations per tree for the proposed (conventional) architecture with $N = 31$ and $K = 256$

OPs per tree	Memory access			Compare	Crossbar
Data	$p_{m,n}$	$c_{m,l}$	$\tau_{m,n}$	$x(p_{m,n}) > \tau_{m,n}$	$x(p_{m,n})$
Bit precision	6(8)	3(3)	8(8)	8(8)	Mux ratio 64 : 1
Size (Bytes)	23.3(31)	0.5(16)	0(31)	1	(256 : 1)
# of OPs[a]	3(4)	0.5(2)	0(4)	31(31)	1(1)

[a]8 bytes per SRAM access

3.3.3 Regularized Trees

The decision trees are extended to form a balanced tree with N nodes (Fig. 3.11) by introducing dummy nodes in order to regularize the memory access and processing patterns. Note that tree regularization does not incur any additional delay overhead or tree node hardware. The reason is that any RF architecture has to accommodate all possible tree shapes, including the worst-case scenario of having to implement

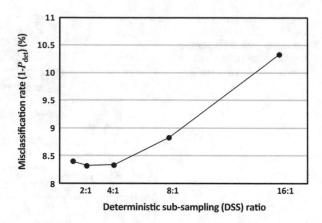

Fig. 3.12 DSS ratio vs. simulated misclassification rate $(1 - P_{\text{det}})$ on the a test dataset containing 700 images

N nodes and incur the worst-case delay of $\log_2(N + 1)$ node comparisons. Furthermore, a regularized tree does not require an additional pointer address for the child node. Though one could choose to disable the dummy nodes selectively to minimize computations in the regularized tree, doing so would require an additional data bit to indicate a null node and an additional memory access. Post-layout simulations show that a 1-b memory access requires 14× higher energy than a comparison operation. Thus, we allow all the nodes to be computed, including dummy nodes. The comparison result $q_{\text{m,n}}$ of the dummy nodes does not affect the classification results, as both left and right paths end up with identical labels. Therefore, for every dummy node, we assign the values of $x(p_{\text{m,n}})$ and $\tau_{\text{m,n}}$ for which $|x(p_{\text{m,n}}) - \tau_{\text{m,n}}|$ is maximized. Doing so minimizes the probability of a metastability event in the analog comparator, as described in Sect. 3.3.5.

3.3.4 Deterministic Sub-sampling (DSS)

The modified RF algorithm (Fig. 3.11) employs a fixed-pattern deterministic sub-sampling (DSS) step prior to RSS to solve the problem of requiring a *complex crossbar*. Figure 3.12 shows that sub-sampling up to a ratio of 4:1 does not affect the misclassification rate because of the highly correlated pixel values. Therefore, a 4:1 DSS ratio is chosen to minimize the crossbar complexity without degrading P_{det}. The optimal DSS ratio depends on the image resolution. The complexity of the subsequent RSS crossbar is reduced from 256:1 to 64:1 via DSS when the input **x** is a 16×16 image ($K = 256$). That reduction has the additional benefit of reducing the precision of $p_{\text{m,n}}$ from 8-b to 6-b.

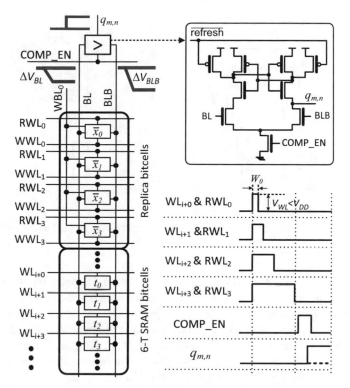

Fig. 3.13 In-memory comparison of $T = \sum_{i=0}^{B-1} 2^i t_i$ with $X = \sum_{i=0}^{B-1} 2^i x_i$ for bit precision $B = 4$

3.3.5 DIMA-Based Comparison

DIMA-based tree processing addresses the requirement of *highly parallel comparisons* (Fig. 3.13) and thus eliminates the need to explicitly fetch the thresholds $\tau_{m,n}$. In-memory comparison requires that the B-b thresholds $\tau_{m,n}$ (T in Fig. 3.13) and indexed pixels $x(p_{m,n})$ (X in Fig. 3.13) be stored in a *column-major* format, i.e., bits of a word are stored in a column. The comparison begins with the simultaneous application of WL access pulses to all the rows storing T and $\overline{X}$ as in the functional read (FR) process. Here, the pulse width is proportional to the bit position so as to generates a BL voltage (discharge) swing ΔV_{BL} that is proportional to $(X - T)$ and is given by:

$$\Delta V_{BL}(T, \overline{X}) = \frac{V_{PRE} W_0}{R_{BL} C_{BL}} \sum_{i=0}^{B-1} 2^i (\overline{t_i} + x_i)$$

$$= \Delta V_{lsb}(X - T - 1) \tag{3.1}$$

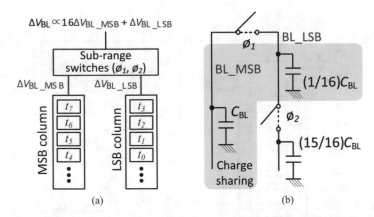

Fig. 3.14 Sub-ranged read [32]: (**a**) column pair implementation, and (**b**) an equivalent capacitance model

where V_{PRE} is the precharged BL voltage, W_0 is the LSB pulse width, R_{BL} is the resistance of the BL discharge path comprising the access and pull-down transistors of the enabled bitcells, and $\Delta V_{lsb} = \frac{V_{PRE}W_0}{R_{BL}C_{BL}}$. The bias '−1' is generated by expressing negative numbers in 2's complement representation [29, 32]. Because of the complementary nature of the SRAM bitcell, ΔV_{BLB} is also proportional to $(T - X)$. Next, BLs and BLBs are fed into analog comparators [66] to generate node-level decisions ($q_{m,n}$) in parallel for all columns.

Integral non-linearity (INL) of FR is improved by up to 65% by sub-ranged read (Fig. 3.14), where $B/2$-b MSBs and $B/2$-b LSBs are read from adjacent columns (from a column pair) followed by a capacitively-weighted charge-sharing step that assigns 16× greater weight to the MSBs [32]. The $(1/16)C_{BL}$ in Fig. 3.14b is implemented by the parasitic capacitance of the sub-range switch and a tunable capacitor to calibrate the accuracy [32]. The WL voltage $V_{WL} < 0.65$ V to prevent destructive read and further improve linearity. The replica BCA [32] (Fig. 3.15) stores $\overline{X}$ via a separate write BL (WBL) and wordline (WWL) to avoid full-swing toggling of high-capacitance BL, as required during the SRAM write operation.

In-memory comparison is a massively parallel operation, as it fetches and processes $B/2$-b per column for each read access while bypassing the column mux, whereas the conventional memory fetches only one bit per L columns per access, where L is the column muxing ratio (with typical values from 4 to 16). FR also saves precharge energy by accessing $B/2$-b per BL precharge, whereas a conventional read fetches a single bit through the column mux per L BL precharges.

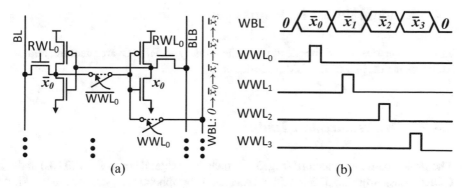

Fig. 3.15 Replica BCA operation [32]: (**a**) bitcell, and (**b**) write access timing diagram

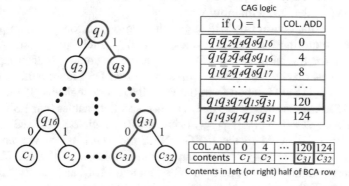

Fig. 3.16 CAG function $g(\ )$, where red tree nodes indicate an example of a selected path. The CAG logic generates the BCA column address corresponding to the chosen label, c_{31}

3.3.6 Class Address (ADD) Generator

The *LUT inefficiency* problem is solved by employing a Class ADD generator (CAG), which converts the result of comparisons $q_{m,n}$s into a memory address for the chosen label $c_{m,l}$. The conversion is achieved by generating the memory address as $f(m) + g(q_{m,1 \sim N})$, where $f(\ )$ provides an offset address and $g(\ )$ specifies the address. The functions $f(\ )$ and $g(\ )$ need to guarantee a one-to-one mapping from $(m, q_{m,n})$ to a memory address. For example, in this paper, $f(\ )$ generates the row address $= 12 \lfloor (m-1)/4 \rfloor + 8 + \mathrm{mod}(\lfloor (m-1)/2 \rfloor$ and decides on either the left or right half of BCA based on $\mathrm{mod}(m+1, 2)$. Function $g(\ )$ specifies the column address via simple Boolean logic with $q_{m,n}$s, as shown in Fig. 3.16. Therefore, only the label of the chosen path (rather than the index and label of every path) is fetched.

These proposed techniques require the fetching of only 24.8 bytes/tree of data rather than 79 bytes/tree as required by the conventional parallel architecture, as summarized in Table 3.6.

3.4   Random Forest IC Prototype

This section describes the design of the prototype RF IC architecture and its implementation in a 65 nm CMOS process.

3.4.1   Architecture and Timing

The prototype IC architecture (Fig. 3.17) includes a digital controller (CTRL) and a CORE consisting of a 512×256 SRAM BCA, peripherals for standard read/write operations, a 64-b I/O ($B_{IO} = 64$) with a 4:1 ($L = 4$) column mux, FR WL drivers, a 4:1 DSS input buffer to store the streamed 256-b $\mathbf{x}$ ($K = 256$), RSS crossbars, a CAG, a label finder, and a majority voter.

The timing diagram in Fig. 3.18 shows that a group of four trees are processed in parallel, requiring 171 clock cycles, and $M/4$ such groups are processed sequentially for a total of $M \leq 168$ trees. In the beginning, four groups of 4:1 sub-sampled 64 pixel words ($\mathbf{x}$) are stored in the four DSS input buffers. First, the pixel indices $p_{m,n}$s are fetched from the BCA into index registers (IREG) through 12 normal SRAM read accesses. Next, a main controller enables four RSS crossbars (CB) that place $B = 8$-b pixels $x(p_{m,n})$s into the RSS registers (RSREG) according to the index $p_{m,n}$ stored in the IREGs. Next, the $x(p_{m,n})$s are written into the replica BCA for in-memory comparison with thresholds $\tau_{m,n}$. The FR WL drivers apply binary weighted PWM pulses to WLs simultaneously, and the discharged BLs and BLBs are fed to analog comparators (Fig. 3.13). The in-memory comparison generates 128 comparison outputs $q_{m,n}$s per precharge cycle at the output of the 128 pitch-matched analog comparators in parallel. The controller fetches four tree-level labels $c_{m,l}$s from the BCA via normal read operation using the address generated by CAG using the in-memory comparison outputs $q_{m,n}$. Finally, the majority voter makes a final decision based on the M tree-level labels $c_{m,l}$ after processing $M/4$ such groups.

3.4.2   Circuit Implementation

The prototype IC (Fig. 3.19) was fabricated in a 65 nm CMOS process and packaged an 80-pin QFN as summarized in Table 3.7. The logic blocks, including the DSS input buffer, IREG, CB, and RSREG occupy, 25% of the area whereas less than 10% of the area is occupied by the additional circuitry for in-memory comparison, such as analog comparators and the replica BCA. The CTRL operates with 1 GHz clock to provide fine time resolution for control signals. In this implementation, all the control signals, even for standard read/write operations, are synced with the clock. On the other hand, the multi-row WL driver in Fig. 3.17 includes a pulse

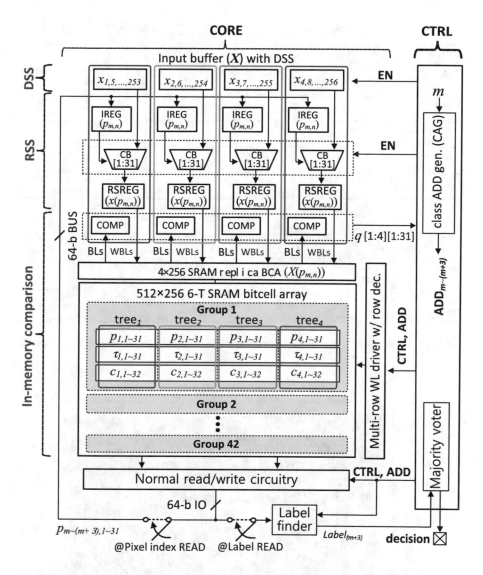

Fig. 3.17 Proposed RF architecture (IREG: pixel index register, RSREG: RSS register, COMP: analog comparator, and CB: crossbar)

generator whose output pulse width is set by the configuration code, which alters the number of inverter-based delay cells.

The 256-b input image **x** is provided serially to the input buffer, and the configuration word defines operations such as the number of trees (M) to be processed. The final 3-b decision, to support a maximum of 8 classes, is fetched through the serial output port.

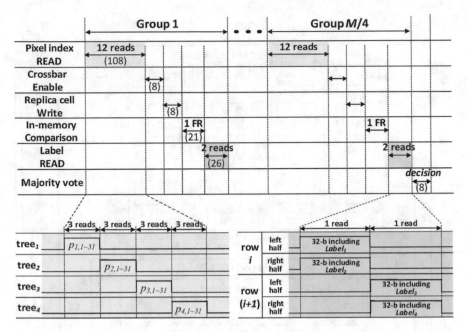

Fig. 3.18 Timing diagram showing the number of required clock cycles per stage and the sequence of operations

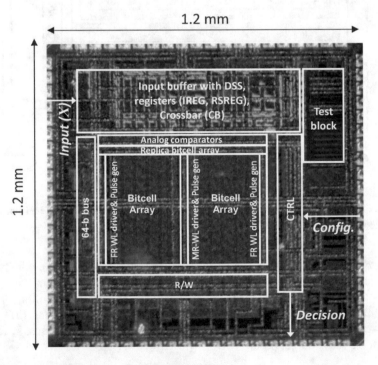

Fig. 3.19 Micrograph of the in-memory RF classifier IC

Table 3.7 Chip summary

Technology	65 nm CMOS
Die size	1.2 mm × 1.2 mm
SRAM capacity	16 kB (512 × 256 bitcells)
Bitcell dimension	2.11 μm × 0.92 μm
CTRL operating freq	1 GHz
Supply voltage (V)	CTRL: 0.75 CORE: 1
Energy per decision (pJ) (4 trees, 64 trees)	CTRL: (0.3,5.0) CORE: (0.9,14.4)
Decision throughput (decisions/s)	4 trees: 5.6 M 64 trees: 364.4 K

3.5 Measured Results

This section describes the measured results from the prototype IC and evaluates its energy, delay, accuracy, and robustness benefits with respect to a conventional system. The conventional system employs the same architecture as in Fig. 3.17, but with a 256:1 RSS crossbar and no DSS, uses a digital comparator (an 8-b subtractor with a sign detector) instead of in-memory comparison, and uses a digital LUT logic to store all the labels $c_{m,l}$s. The digital comparators process eight comparisons at a time with a 64-b output (eight $\tau_{m,n}$s) from the SRAM, while the next eight words are fetched. The energy and delay of the conventional architecture are estimated from: (1) measurements of the normal SRAM read access of the prototype IC, and (2) post-layout simulations of synthesized digital comparators and a 256:1 crossbar. The required operations (memory access, comparison, and enabling crossbar) of the conventional architecture are calculated using Table 3.6.

Layouts of the digital comparator and the 256:1 crossbar were matched to the horizontal dimension of the SRAM BCA to align well with the SRAM IOs. The area of the synthesized 256:1 RSS crossbar was found to be four times larger than that of the proposed crossbar because of its higher complexity, whereas the area of the digital comparators was similar to that of the proposed analog processors (i.e., the replica BCA and analog comparators). Therefore, the conventional architecture occupies approximately 1.8× more area than DIMA.

3.5.1 Application Mapping and Classifier Training

The proposed architecture was tested on two datasets (Fig. 3.20): (a) the KUL Belgium traffic sign dataset [67] for an 8-class traffic sign recognition task, and (b) the MIT CBCL dataset [62] for a face detection task. The input pixels and the threshold values $\tau_{m,n}$ were represented in 8-b fixed-point as this precision provides approximately the same accuracy as floating point implementation [1, 39, 59].

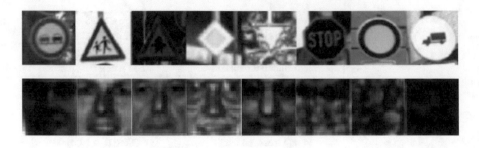

Task	Class	Dataset	Image Size
Traffic sign recognition [67]	8	KUL Belgium traffic sign dataset - Train: 148 images per class - Test: 200 random images	Resized 16×16 pixels (gray-scale)
Face detection [62]	2	MIT CBCL dataset - Train: 2000 images per class - Test: 200 random images	

Fig. 3.20 Summary of the datasets used in measurements

During off-chip training, 148 training images per class were used for traffic sign recognition whereas 2000 training images per class were used for face detection. The maximum supported tree depth was chosen to be six, which is considered optimal for the target application [68]. To evaluate the impact of the number of trees on accuracy, two different cases of $M = 4$ and $M = 64$ were tested. The classification accuracy (P_{det}) was measured by streaming 200 randomly chosen images and counting the correct decisions.

3.5.2   Component-Level Accuracy

Figure 3.21 shows that in-memory subtraction, which is realized as part of the FR process, achieves an INL < 1.85 LSB in the range of $-15 \leq T_{MSB} - X_{MSB} \leq 15$. Deviations in V_{BL} are < 25 mV over different combinations of T_{MSB} and X_{MSB}. Those variations are induced by circuit non-idealities, including inaccurate ratios of PWM pulse widths, asymmetries of replica BCAs, and the BL voltage dependence of the discharge path resistance. Spatial transistor threshold voltage (V_t) variations caused by random dopant fluctuations lead to increased comparator errors, as shown in Fig. 3.22, where the in-memory comparator accuracy ranges from 100% to 50% (when $T_{MSB} \simeq X_{MSB}$). The asymmetry of the replica BCA is due to the use of single-ended WBL, as shown in Fig. 3.15, which results in an asymmetric discharge current between BL and BLB. That asymmetry leads to an increase in the

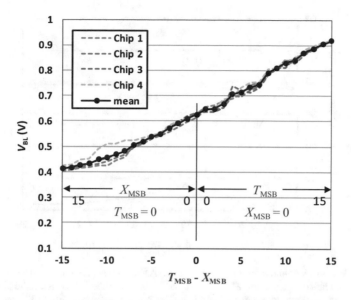

Fig. 3.21 Measured accuracy of in-memory subtraction from four dies where T_{MSB} and X_{MSB} are the decimal representations of the top four MSBs of the threshold and input pixel, respectively

Fig. 3.22 Measured accuracy of DIMA comparisons with all possible combinations of (X_{MSB}, T_{MSB}) with $\Delta V_{lsb} =$ 25 mV. Each datapoint was obtained by averaging 256 measurements over 256 different locations of the BCA

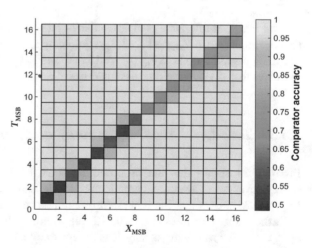

comparator error rate for large values of X_{MSB} and T_{MSB}, e.g., $X_{MSB} > 10$ and $T_{MSB} > 10$.

Figure 3.23 shows that the measured in-memory comparison error rate increases from 1.6% to 14.5% because of the increased impact of process variations as ΔV_{lsb} (see (3.1)) reduces from 25 mV to 5 mV. Comparator errors were measured at each ΔV_{lsb} by counting the errors during the classification with the KUL Belgium dataset.

The impact of in-memory comparison errors on the misclassification rate was studied next. System simulations indicate that a comparison error rate of less than 9.5% will result in an indiscernible 8-class classification accuracy loss, whereas

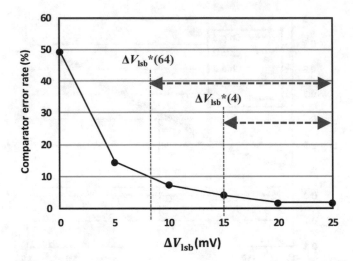

Fig. 3.23 Measured comparator error rate. Here, $\Delta V_{\text{lsb}}^*(M)$ is the minimum ΔV_{lsb} to avoid accuracy degradation when using M trees. The ΔV_{lsb} is controlled by changing the voltage level of WL, enabling a signal to affect R_{BL} in (3.1). The ΔV_{lsb} is estimated from (3.1) by measuring ΔV_{BL} with $X = 15$ and $T = 0$ in the test mode

$M = 4$ trees can tolerate a comparison error rate of only 4%. Thus, Fig. 3.23 shows that the minimum ΔV_{lsb} ($\Delta V_{\text{lsb}}^*(M)$) required by an M-tree RF architecture to avoid accuracy degradation is $\Delta V_{\text{lsb}}^*(4) = 15\,\text{mV}$, and $\Delta V_{\text{lsb}}^*(64) = 8\,\text{mV}$. This analysis indicates that ΔV_{lsb} should be assigned based on the number of trees (M).

3.5.3   Task-Level Accuracy, Energy, and Throughput

Figure 3.24 shows that the proposed IC achieves a maximum of 3.1× energy savings over the conventional architecture for the same misclassification rate of 6%. Figure 3.25 shows that the inherent error tolerance of the RF algorithm improves with the number of trees (M). The RF algorithm with $M = 64$ trees can operate at ΔV_{lsb} of 15 mV without any loss in accuracy as compared to an ideal fixed-point implementation (Fig. 3.25) while the RF with four trees observes degradation as soon as $\Delta V_{\text{lsb}} < 20\,\text{mV}$ for both datasets. This requirement is 5–7 mV greater than that predicted by Fig. 3.23 where only errors from FR and analog comparators were taken into account. The discrepancy can be attributed to the presence of additional error sources in other stages, such as the pixel read and label read, where conventional SRAM read may also become erroneous due to ΔV_{lsb} scaling. It is expected the energy efficiency will improve in highly complex real-world tasks with a few hundreds trees, as it will enable further reductions in ΔV_{lsb}. Binary face detection is less sensitive to ΔV_{lsb} reduction than 8-class traffic sign recognition is as

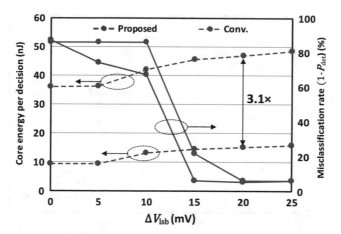

Fig. 3.24 Energy vs. misclassification rate with regard to ΔV_{lsb} for face detection with $M = 64$ trees, where $\Delta V_{BL} = 8\Delta V_{lsb}$ during a normal SRAM read in order to achieve a zero bit-error rate at the default configuration ($\Delta V_{BL}= 200$ mV, $\Delta V_{lsb}= 25$ mV)

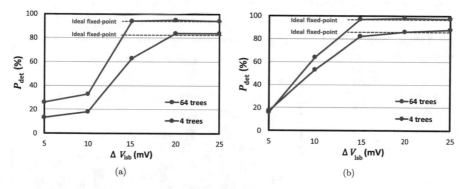

Fig. 3.25 Task-level accuracy vs. ΔV_{lsb} for different numbers of trees (M): (**a**) traffic sign recognition, and (**b**) face detection

expected. That result indicates that the ΔV_{lsb} can be systematically scaled based on the number of classes, the number of trees M, and the target classification accuracy.

Figure 3.26 shows the energy vs. accuracy trade-off with respect to two parameters: (1) the number of trees (M), and (2) the BL voltage swing ΔV_{lsb}. The decision energy can be saved by reducing either M or ΔV_{lsb}. However, the P_{det} degrades more gracefully when the energy is reduced by reducing M, e.g., P_{det} drops by 10%, achieving 14.5× energy savings with M, whereas P_{det} degrades by 68% for 1.7× energy reduction with ΔV_{lsb}. Therefore, classification with smaller M always achieves better energy efficiency for a fixed classification accuracy P_{det}, though the maximum achievable P_{det} also reduces.

To observe the impact of process variations, P_{det} is measured by testing five dies. Figure 3.27 shows minor differences in P_{det}, e.g., <6% and <2% with $M = 4$ and

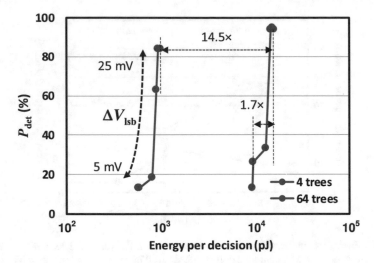

Fig. 3.26 Energy vs. task-level accuracy for traffic sign recognition with different numbers of trees (M) at $\Delta V_{lsb} = 5 - 25 \, \text{mV}$

Fig. 3.27 Measured misclassification rates of multiple chip instances for traffic sign recognition with different numbers of trees (M) at $\Delta V_{lsb} = 25 \, \text{mV}$

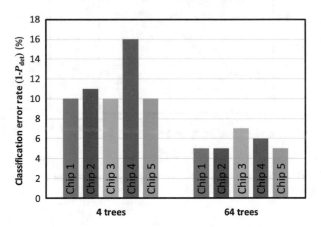

64, respectively. This result indicates that the inherent error tolerance of ensemble classification in the RF algorithm compensates for the process variations.

Figure 3.28 shows the energy and delay breakdowns of the proposed IC as compared to the conventional architecture. The breakdowns were obtained from post-layout simulations as it was difficult to measure the energy and delay of each component from measurement of the prototype IC. The energy and delay were reduced by 3.1× and 2.2×, respectively, thereby providing an overall 6.8× lower energy-delay product (EDP). Figure 3.28 indicates that DSS, in-memory

Fig. 3.28 Energy and delay breakdown and the comparison with the conventional digital architecture at $\Delta V_{lsb} = 25\,mV$ ($\Delta V_{BL} = 200\,mV$) obtained via post-layout simulations

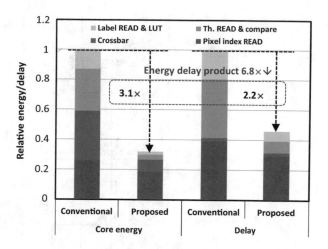

comparison, and CAG contribute a 47%, 37%, and 16% of the total energy savings, and 18%, 58%, and 24% of the total delay reduction, respectively, thereby indicating the relative contributions of DIMA and non-DIMA techniques.

Table 3.8 shows that the prototype IC achieved a throughput of 364 K decisions/s and energy efficiency of 19.4 nJ/decision including the energy from CTRL. To the best of our knowledge, this is the first IC prototype of the RF algorithm. Therefore, this IC is compared with other tree-based classifier ICs [69, 70]. Because of the vast difference in the datasets, tree complexities, and node architectures, focus is on node-level metrics (the last three columns in Table 3.8). The proposed RF classifier achieves a node-level EDP that is three-to-four orders of magnitude lower than those in [69, 70].

3.6 Conclusion

This chapter described two DIMA prototype ICs: (1) a multi-functional DIMA for SVM, TM, k-NN, and MF, and (2) a DIMA for the RF algorithm. Potentially, more algorithms can be covered by simply modifying or adding functionality in each processing stage. Measurement results for the multi-functional DIMA IC demonstrates up to 53× EDP reduction relative to the conventional digital architecture optimally designed for each algorithm. Furthermore, the EDP benefit is expected to be even higher (up to 56×) in multi-bank scenarios because the controller overhead is shared over many banks. The prototype IC of RF also achieves a 3.1× energy savings and 2.2× speed-up while providing a 6.8× lower EDP relative to the conventional digital architecture while having the same accuracy of >93%, thereby leading to a throughput of 364 K decisions/s and energy efficiency of 19.4 nJ/decision for the 8-class traffic sign recognition problem. A trade-off

Table 3.8 Comparison of energy efficiency and throughput with prior art

Prior art	Process	Algorithm	Power (mW)	Layout area (mm²)	Decision throughput (decisions/s) ①	Decision energy (nJ/decision) ②	# of scalar node computations /decision ③ᵃ	Node energy (pJ) ④ = ②/③	Decision EDP (f J·s/decision) ⑤ = ②/①	Node EDP (10⁻¹⁷J·s) ⑤/③
[69]	65 nm CMOS	Vocabulary tree	5.6	6.3	30	186.7K	8.4M	22	6.2G	74K
[70]	65 nm CMOS	Vocabulary forest	27.6	2.3	60	460K	320M	1.4	7.7G	2.4K
This workᵇ	65 nm CMOS	Random forest	7.1	1.0	364.4K	19.4	1984	9.8	53.2	2.7

ᵃ(# of trees/decision) × (# of tree nodes/tree) × (dimension/tree node), e.g., 64 × 31 × 1 in this work
ᵇAt $\Delta V_{\text{lsb}} = 15$ mV with $M = 64$ trees and misclassification rate of 94%

between the energy and accuracy was also observed in both ICs by controlling the ΔV_{BL}. This indicates that there is a potential to push DIMA's energy savings further through the use of statistical error compensation techniques. Specifically, the ML coefficients can be re-trained to compensate the deterministic error patterns of non-ideal analog circuity. On the other hand, DIMA with an on-chip trainer will be able to compensate time-dependent noise sources. In particular, the feasibility of multi-functional DIMA indicates that there is potential to realize a programmable DIMA instruction set architecture (ISA), which will be explored in Chap. 6.

Chapter 4
A Variation-Tolerant DIMA via On-Chip Training

Though Chaps. 2 and 3 have shown that DIMA significantly reduces the decision EDP of machine learning kernels, it is not clear what the limits of this reduction are. In fact, DIMA's analog nature makes it sensitive to process, voltage, and temperature (PVT) variations especially when the BL swing is reduced in order to enhance DIMA's energy efficiency. This chapter describes the use of an on-chip trainer to reduce the BL swing without compromising DIMA's inference accuracy. A robust deep in-memory SVM classifier prototype in a 65 nm CMOS that uses a standard 16 kB 6T SRAM is presented. This IC employs an on-chip stochastic gradient descent (SGD)-based trainer that adapts not only to chip-specific variations in PVT but also to data statistics in order to further enhance DIMA's energy efficiency.

4.1 Background and Rationale

4.1.1 SGD-Based SVM Classifier

The SVM is a maximum-margin binary classifier that predicts the class label $y_k \in \{\pm 1\}$ as follows:

$$\mathbf{w}^T \mathbf{x}_k + b \underset{\hat{y}_k = -1}{\overset{\hat{y}_k = +1}{\gtrless}} 0 \tag{4.1}$$

where $\hat{y}_k$ is the predicted kth class label, and the D-dimensional vectors $\mathbf{x}_k = [X_{0,k}, X_{2,k}, \ldots, X_{D-1,k}]^T$, weights $\mathbf{w} = [W_0, W_2, \ldots, W_{D-1}]^T$, and a scalar bias b are referred to as the parameters of the SVM algorithm. The SVM cost function per sample $Q(\mathbf{w}, \mathbf{x}_k)$ is given by Bottou et al. [71],

© Springer Nature Switzerland AG 2020
M. Kang et al., *Deep In-memory Architectures for Machine Learning*,
https://doi.org/10.1007/978-3-030-35971-3_4

$$Q(\mathbf{w}, \mathbf{x}_k) = \frac{\lambda}{2}(\|\mathbf{w}\|_2^2 + b^2) + [1 - y_k(\mathbf{w}^T \mathbf{x}_k + b)]_+ \qquad (4.2)$$

where λ is a regularization factor, y_k is the true label for the data sample $\mathbf{x}_k$, and $[x]_+ = \max\{0, x\}$ is the so-called hinge loss.

Stochastic gradient descent (SGD) has emerged as one of the most effective training algorithms for machine learning [71] that includes SVM [72]. The SGD algorithm minimizes the cost function $Q(\mathbf{w}, \mathbf{x}_k)$ averaged over the training set by updating the weight vector iteratively, as follows:

$$\mathbf{w}_{m+1} = (1 - \lambda\gamma)\mathbf{w}_m + \gamma\frac{1}{N}\sum_{n=0}^{N-1}\begin{cases}0 & \text{if } y_n^{(m)}(\mathbf{w}_m^T\mathbf{x}_n^{(m)} + b_m) > 1 \\ y_n^{(m)}\mathbf{x}_n^{(m)} & \text{otherwise}\end{cases}$$

$$(4.3)$$

$$b_{m+1} = (1 - \lambda\gamma)b_m + \gamma\frac{1}{N}\sum_{n=0}^{N-1}\begin{cases}0 & \text{if } y_n^{(m)}(\mathbf{w}_m^T\mathbf{x}_n^{(m)} + b_m) > 1 \\ y_n^{(m)} & \text{otherwise}\end{cases} \qquad (4.4)$$

where N is the batch size, m is the batch index, $\mathbf{w}_m$ is the weight vector in the mth batch, $\mathbf{x}_n^{(m)}$ is the nth sample of the mth batch, and γ is the learning rate.

The primary inference computation in the SVM algorithm is the dot product $\mathbf{w}^T\mathbf{x}$. For an SVM implementation on DIMA, the weights $\mathbf{w}$ are stored in the BCA and accessed a functional read, while the BLPs execute element-wise multiplication of $\mathbf{w}$ and $\mathbf{x}$, and the CBLP aggregates the BLP outputs to obtain the final dot product.

4.1.2 A Systems Rationale for On-Chip Training

This section provides a systems rationale for using the SGD algorithm on-chip to compensate for PVT variations in DIMA. Those variations are caused by Kang et al. [32]: (1) spatial transistor threshold voltage (V_t) variations caused by random dopant fluctuations [29]; (2) dependence of the discharge path (access and pull-down transistors in the bitcell) current on the BL voltage; and (3) the finite transition (rise and fall) times of the PWM WL pulses. As a result, the functional read step of DIMA, which accesses weights W_j stored in the jth column, generates a BL discharge $\Delta V_{\text{BL},j} = \alpha_j W_j$ on the jth BL. Therefore, even though the BCA stores $\mathbf{w} = [W_0, W_1, \ldots, W_{D-1}]$, DIMA implements the dot product in (4.1) with per-dimension scaled weights $\mathbf{w}' = [\beta_0 W_0, \beta_1 W_1, \ldots, \beta_{D-1}W_{D-1}]$, where $\beta_j = \frac{\alpha_j}{\alpha}$ (with $\alpha_j = \alpha(1 + \frac{\Delta\alpha_j}{\alpha})$) is the per-dimension proportionality factor. Measured results show that this simple model captures the effects of spatial PVT variations to a first order.

Thus, the SGD algorithm minimizes the modified cost function $Q'(\mathbf{w}, \mathbf{x}_k)$ given by:

$$Q'(\mathbf{w}, \mathbf{x}_k) = \frac{\lambda}{2}(||\mathbf{w}'||^2 + b^2) + [1 - y(\mathbf{w}'^T \mathbf{x}_k + b)]_+$$

$$= \frac{\lambda}{2}\left(\sum_{j=0}^{D-1}\beta_j^2 W_j^2 + b^2\right) + \left[1 - y\left(\sum_{j=0}^{D-1}\beta_j W_j X_{j,k} + b\right)\right]_+ \tag{4.5}$$

The modified cost function $Q'(\mathbf{w}, \mathbf{x})$ can be shown to be convex in $\mathbf{w}$, which implies that the SGD algorithm will converge to the global minimum in the presence of PVT variations. However, as the proportionality factor β_j is unknown in practice and is die-specific, we employ a per-dimension learning rate of $\gamma_j = \gamma/\beta_j$ is employed to obtain the following SGD update:

$$\mathbf{w}_{m+1} = \mathbf{w}_m - \lambda\Gamma\mathbf{w}_m$$

$$+ \Gamma\frac{1}{N}\sum_{n=0}^{N-1}\begin{cases} 0 & \text{if } y_n^{(m)}(\mathbf{w}_m'^T \mathbf{x}_n^{(m)} + b_m) > 1 \\ y_n^{(m)}\mathbf{x}_n^{(m)} & \text{otherwise} \end{cases} \tag{4.6}$$

which is identical to (4.3) except that it relies upon computation of the dot product in the presence of spatial variations, where $\Gamma = \text{diag}(\gamma_0, \ldots, \gamma_{D-1})$ is the per-dimensions learning rate.

One might consider using on-chip training in a digital architecture to compensate for PVT variations under reduced ΔV_{BL}. However, that method would be ineffective in such architectures because PVT variations with reduced ΔV_{BL} lead to increased sense amplifier bit errors, including most-significant-bit (MSB) errors. These large-magnitude errors in turn lead to a large increase in the misclassification rate. Figure 4.1 shows that on-chip learning cannot compensate for the increase in the misclassification rate due to reduced ΔV_{BL}. DIMA avoids such errors by reading a weighted function of the bits of W instead of reading the bits directly.

4.2 Architecture and Circuit Implementation

The architecture of the prototype IC in Fig. 4.2 has four major blocks: (a) the in-memory CORE (IM-CORE) to execute the inference computations; (b) the standard read/write SRAM interface; (c) the digital trainer; and (d) a digital controller (CTRL) to sequence the operations. The prototype IC has three modes of operation: (1) standard SRAM; (2) in-memory inference; and (3) on-chip training.

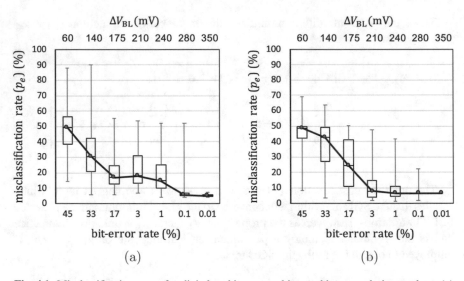

Fig. 4.1 Misclassification rate of a digital architecture subject to bit errors during readout: (**a**) before retraining, and (**b**) after retraining. The bit error rate (BER) is obtained via measurements from the prototype IC under reduced BL swing (ΔV_{BL}) in the SRAM mode. The misclassification rate is obtained via simulations of 1000 independent instances of an SRAM bank, with each instance processing 858 data samples, in the presence of randomly assigned bit-flip locations, at a rate based on the BER

4.2.1   The IM-CORE Block

The in-memory inference mode is implemented via the IM-CORE block. The IM-CORE comprises a conventional 512×256 6T SRAM BCA and in-memory computation circuitry, which includes: (a) pulse-width -modulated WL drivers to realize functional read; (b) BLPs implementing signed multiplication; (c) CBLP implementing summation; (d) an ADC bank; and (e) a comparator bank to generate final decisions. Though the 16-b SVM weights **w** are stored in the BCA, only 8-b MSBs of the weights **w** are used during inference (see Fig. 4.3). Functional read simultaneously generates ΔV_{BL} discharge voltages on all 256 BLs in one read cycle. The 8-b input samples **x** are streamed into the input buffer, and are transferred to the BLPs via a 256-b bus in four cycles for element-wise multiplication of **w** and **x**, which are then summed in the CBLP to obtain the dot product in (4.1).

Functional read performs digital-to-analog conversion of the weights stored in the SRAM array such that the discharge on the BL represents the analog values of the weights being read.

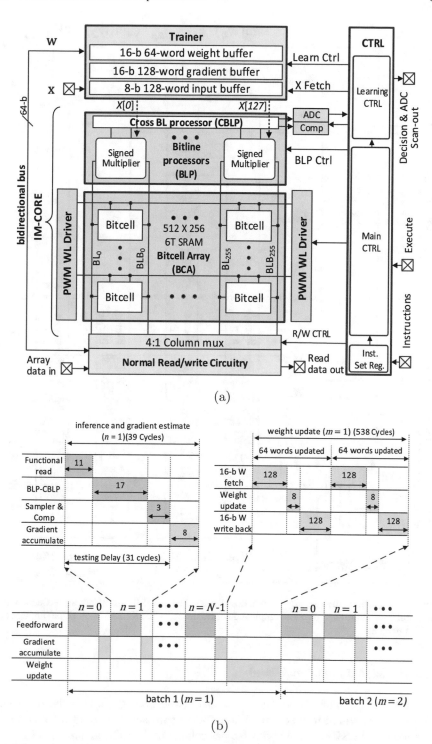

Fig. 4.2 Proposed system: (**a**) chip architecture showing the IM-CORE, trainer and CTRL, and (**b**) the timing diagram

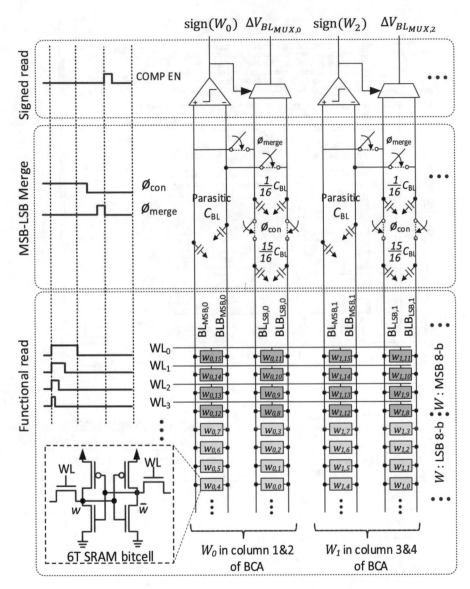

Fig. 4.3 Implementation of signed functional read. The 8 MSBs of W are stored across two adjacent bitcell columns (red) and employed for inference, while 8 LSBs of W (blue) are updated during training

4.2.1.1   Signed Functional Read

The 8-b SVM weights $W \equiv \{w_7, .., w_1, w_0\}$ (where $w_i \in \{0, 1\}$ is the ith binary bit of W) are stored in a column-major format split across two adjacent columns (the MSB and LSB columns) with 4-b per column in the BCA as shown in Fig. 4.3.

The application of WL access pulses with binary weighted pulse widths (PWM) followed by an LSB-MSB merge results in discharge voltages on BL and BLB as follows:

$$\Delta V_{\mathrm{BL}} = \frac{V_{\mathrm{PRE}}T_0}{R_{\mathrm{BL}}C_{\mathrm{BL}}} \left[\frac{1}{16} \sum_{i=0}^{3} 2^i \overline{w}_i + \sum_{i=4}^{7} 2^i \overline{w}_i \right] \tag{4.7}$$

$$\Delta V_{\mathrm{BLB}} = \frac{V_{\mathrm{PRE}}T_0}{R_{\mathrm{BL}}C_{\mathrm{BL}}} \left[\frac{1}{16} \sum_{i=0}^{3} 2^i w_i + \sum_{i=4}^{7} 2^i w_i \right] \tag{4.8}$$

where V_{PRE} is the BL precharge voltage, T_0 is the least significant bit (LSB) pulse width, C_{BL} is the BL capacitance, and R_{BL} is the BL discharge path resistance of a bitcell. A two's complement representation for W is employed to account for its sign. The magnitude $|W|$ is obtained by choosing ΔV_{BLB} (ΔV_{BLB}) if W is negative (positive) since the two discharges are complementary. The sign of W is obtained by comparing ΔV_{BLB} and ΔV_{BL} as shown in Fig. 4.3. Thus, the discharge on the output node of functional read $\Delta V_{\mathrm{BL_{MUX}}} \propto |W|$. The detected sign and the magnitude are then passed on to the signed multipliers in the BLP.

Equation (4.8) assumes that R_{BL} is invariant with respect to the time-varying ΔV_{BL} and spatially across the BCA. In practice, the amplitude of word-line voltages V_{WL} is kept sufficiently low (0.45–0.65 V) to alleviate the impact of ΔV_{BL} on R_{BL}. Doing so results in the integral non-linearity (INL) of ΔV_{BLB} to be less than 0.65 LSB (see Fig. 4.4a).

Spatial variations in the access transistor threshold voltages lead to variations in R_{BL}, and hence to variations in ΔV_{BL} across the BCA columns. Figure 4.4b shows that the variance of ΔV_{BLB} at $\Delta V_{\mathrm{BL,max}} = 320\,\mathrm{mV}$ is 60% higher than at $\Delta V_{\mathrm{BL,max}} = 440\,\mathrm{mV}$. Figure 4.4c shows that the misclassification rate $p_e = 4\%$ at $\Delta V_{\mathrm{BL,max}} = 560\,\mathrm{mV}$ is within 1% of the floating point accuracy. However, p_e increases to 16% when $\Delta V_{\mathrm{BL,max}}$ is reduced to 320 mV illustrating the impact of spatial variations in ΔV_{BL} on the accuracy of inference. One can reduce the impact of spatial variations on ΔV_{BL} by increasing the maximum BL discharge voltage $\Delta V_{\mathrm{BL,max}}$ (ΔV_{BLB} when the 4-b $W = 15$) by tuning the WL voltage V_{WL} and T_0.

That sensitivity to chip-specific spatial process variations indicates the need to explore on-chip compensation methods. In addition, increasing $\Delta V_{\mathrm{BL,max}}$ to reduce the impact of spatial variations incurs an energy cost leading to an interesting trade-off between computational accuracy and energy in DIMA that can be exploited at the architectural level.

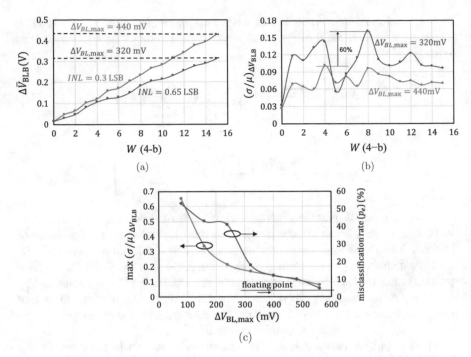

Fig. 4.4 Impact of spatial variations on functional read obtained by measurements across 30 randomly chosen 4-row groups: (**a**) average $\Delta \overline{V}_{BLB}$, (**b**) normalized variance $(\sigma/\mu)_{\Delta V_{BLB}}$, and (**c**) the impact of spatial variations on $(\sigma/\mu)_{\Delta V_{BLB}}$ and misclassification rate p_e with respect to $\Delta V_{BL,max}$

4.2.1.2   BLP: Signed Multiplication

The BLP block needs to implement a signed 8-b×8-b element-wise product $(W_i X_i)$ between $\mathbf{w}$ and $\mathbf{x}$. The BLP block realizes that by using an MSB and LSB multiplier pair with each multiplying the functional read output $\Delta V_{BL_{MUX}} \propto$ 8-b $|W|$ with four MSBs (X_{MSB}) and four LSBs (X_{LSB}) of an 8-b input X. The proposed charge-domain signed multiplier in Fig. 4.5 is based on the unsigned version in [32].

The LSB multiplier receives digital inputs $X_{LSB} \equiv \{x_3, .., x_0\}$ (where $x_i \in \{0, 1\}$ is ith bit of X), the analog $\Delta V_{BL_{MUX}}$, and a digital sign(W) from functional read. Multiplication occurs in four phases, P1–P4, with P3 and P4 overlapping in time (see Fig. 4.5a). The multiplier (Fig. 4.5b) employs six equally sized 25 fF capacitors $(C_0, C_1, C_2, C_x, C_p,$ and $C_n)$, which are initialized to V_{PRE} and either C_p or C_n is chosen as the output capacitor based on sign(W) (P1). Next, the five capacitors nodes are charge-shared with node $\Delta V_{BL_{MUX}}$ using the $\phi_{3,i}$ and ϕ_{dump} switches (P2). Next, the capacitors C_i to V_{PRE} are conditionally charged using the $\phi_{2,i}$ switches (P3). Capacitors $C_x, C_0, C_1,$ and $C_2,$ are sequentially charge shared using the $\phi_{3,i}$ switches (P4), resulting in:

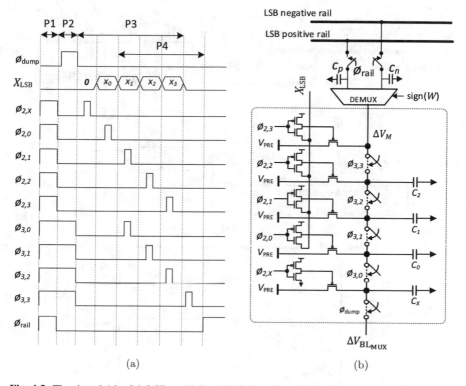

Fig. 4.5 The signed 4-b×8-b LSB multiplier: (**a**) timing diagram, and (**b**) circuit schematic

$$\Delta V_{\mathrm{M}} = \frac{\Delta V_{\mathrm{BL_{MUX}}}}{16} \sum_{i=0}^{3} 2^i x_i \propto |X_{\mathrm{LSB}}||W| \qquad (4.9)$$

where ΔV_{M} is the discharge on the output capacitors C_p or C_n connected to the positive and negative output rails, respectively (see Fig. 4.5b). The MSB multiplier operates identically.

4.2.1.3 Aggregation and Final Decision

The CBLP aggregates the outputs of the MSB and LSB BLP multipliers from across the BCA in order to generate the dot product $\mathbf{w}^T\mathbf{x}$. The aggregation is accomplished (Fig. 4.6a) through merging of the positive (negative) rails from both the MSB and LSB multipliers across the BCA with a 16:1 charge-sharing ratio (see Fig. 4.6a). That merging step results in the voltages $\Delta V_{\mathrm{S},p}$ and $\Delta V_{\mathrm{S},n}$ representing the magnitudes of the sums of the positive and negative terms, respectively, in the final dot product. Therefore, the dot product is computed as:

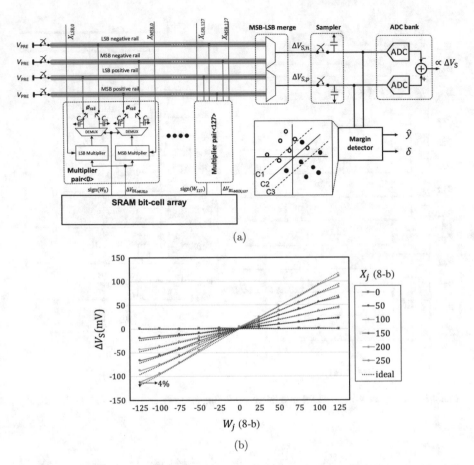

Fig. 4.6 Aggregation in the CBLP: (**a**) circuit schematic, and (**b**) measured output using identical values of W_j and X_j across all the BCA columns, at $\Delta V_{BL,max} = 320\,mV$

$$\Delta V_S = \Delta V_{S,p} - \Delta V_{S,n} \propto \mathbf{w}^\mathsf{T}\mathbf{x} \tag{4.10}$$

The voltages $V_{S,p}$ and $V_{S,n}$ are sampled and processed by a bank of three comparators (C1–C3) where C2 generates the predicted label ($\hat{y}$) while C1 and C3 realize the SVM *margin detector* (see Sect. 4.2.2). The sampled rail voltages $V_{S,p}$ and $V_{S,n}$ are also converted to digital via a 6-b ADC pair for testing purposes. Measured values of ΔV_s (Fig. 4.6b) were found to lie within $< 4\%$ of the dynamic range when $\Delta V_{BL,max} = 320\,mV$.

4.2.2   Trainer

Since the trainer is implemented in digital and the energy and latency cost of SRAM writes are high, the updated weights are written once per batch into the BCA. The trainer (see Fig. 4.2) implements a reformulated version of the batch update (4.6) shown below:

$$\mathbf{w}_{m+1} = (1 - \gamma\lambda)\mathbf{w}_m + \frac{\gamma}{N}\Delta_{W,N}^{(m)} \tag{4.11}$$

where $\Delta_{W,N}^{(m)}$ is the *batch gradient estimate* generated by accumulating the *sample gradient estimate* $\Delta_{W,n}^{(m)}$:

$$\Delta_{W,n+1}^{(m)} = \Delta_{W,n}^{(m)} + \begin{cases} y_n^{(m)}\mathbf{x}_n^{(m)} & \text{if } \delta_n^{(m)} \leq 0 \\ 0 & \text{otherwise} \end{cases} \tag{4.12}$$

where $y_n^{(m)}$ is the true label corresponding to the data sample $\mathbf{x}_n^{(m)}$, and $\delta_n^{(m)}$ is computed in the margin detector (see Fig. 4.6a) as follows:

$$\delta_n^{(m)} = y_n^{(m)}[\text{sign}(z_n^{(m)} - 1) + \text{sign}(z_n^{(m)} + 1)] \tag{4.13}$$

and $z_n^{(m)} = \mathbf{w}_m^T\mathbf{x}_n^{(m)} + b$ is the output of the CBLP block. The margin detector also generates the SVM decision $\hat{y}_n^{(m)} = \text{sign}(z_n^{(m)})$.

The trainer consists of an input buffer to store the streamed input $\mathbf{x}$ and a gradient buffer to store $\Delta_{W,n}$. The trainer implements (4.11) and (4.12) by reusing 64 16-b adders to conserve area. The weight update is performed in two cycles, where 64 words of $\mathbf{w}$ are updated per cycle (see Fig. 4.2b). Multiplication with γ/N and $\gamma\lambda$ in (4.11) are implemented using barrel shifters, which restricts them to powers of 2 in the range $[1, 2^{-15}]$. The ability to choose learning rates in powers-of-2 provides a wider tuning range than one can obtain with learning rates chosen on a linear scale. A wider tuning range in fact allows one to operate with very small learning rates, thereby enabling fine-tuning of the weights to achieve a lower misclassification rate.

4.2.2.1   Trainer Precision Assignment

The precision of the trainer needs to be chosen carefully in order to minimize the cost of training without compromising the convergence behavior of the SGD algorithm. The minimum bit precision for W and $\Delta_{W,n}$ needs to be set. Additionally, the minimum precision for W during inference (4.1) and training (4.11), denoted by B_W and B_{WUD}, respectively, can be substantially different [73]. To avoid gradient accumulator overflow in (4.12), the precision of $\Delta_{W,n}$ (B_Δ) is bounded by:

$$B_\Delta \geq B_x + \log_2 N \tag{4.14}$$

where B_x is the precision of X, which is fixed at 8-b in this application. The precision $B_\Delta = 16$ is set as $B_x = 8$ and batch sizes of up to $N = 256$ were to be accommodated.

It can be shown [73] that a necessary condition for convergence (stopping criterion) of the SGD update in (4.11) is given by,

$$B_{\text{WUD}} \geq 1 - \log_2 \gamma \tag{4.15}$$

The precision $B_{\text{WUD}} = 16$ is set as algorithmic simulations indicate that the algorithm converges with a learning rate $\gamma \geq 2^{-15}$. In addition, the batch-mode algorithm offers an interesting trade-off between the batch size N and the learning rate γ, which will be studied in Sect. 4.3.2.

The regularization factor λ has an optimum value that lies in between an upper bound that constrains the magnitude of W and a lower bound needed to avoid overflow in the weight accumulator (4.11). It can be shown that a sufficient condition to prevent overflow in the weight accumulator is:

$$\lambda \geq \Pr\{y(\mathbf{w}_{\text{opt}}'^T \mathbf{x} + b) < 1\} \tag{4.16}$$

where $\mathbf{w}_{\text{opt}}'$ are the optimal weights after full convergence. Similarly, there exists an upper bound on λ beyond which $|W|$ gets constrained so heavily that the MSBs are forced to zero.

4.3   Experimental Results

This section describes the measured results from the prototype IC and evaluates the effectiveness of on-chip learning in enhancing robustness. The 65 nm CMOS prototype IC (see Fig. 4.7) with a 16 kB SRAM is packaged in an 80-pin QFN. The area overhead of the in-memory computation circuits and the trainer are 15% and 35% of the IM-CORE area, respectively. The overhead of the trainer stems from the need to store intermediate gradients $\Delta_{W,n}$.

4.3.1   Training Procedure

The prototype IC is evaluated on the MIT CBCL face detection dataset [62]. The task is a binary classification problem in which the dataset consists of a separate sets of 4000 training and 858 test images, with equal number of *face* and *non-face*

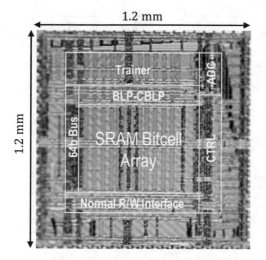

Technology	65nm CMOS	
Die size	1.2 mm × 1.2mm	
Memory capacity	16KB (512 × 256)	
Nominal supply	1.0 V	
CTRL operating frequency	1 GHz	
Energy per decision (nJ)	Test	0.21
	Training	0.34
Average throughput (decision/s)	Test	32.3 M
	Training	21 M

Fig. 4.7 Chip micrograph and summary

images. The input images were scaled down in size to 11×11 pixels and then extended to accommodate the bias term b so that an 8-b 128-dimensional weight vector can be stored in four rows of the BCA.

During training, the batch $\{\mathbf{x}_n^{(m)}\}_{n=0}^{N-1}$ was generated from the training set by random sampling with replacement. During convergence, at the end of every 8th batch, the misclassification rate $p_e = \Pr\{y \neq \hat{y}\}$ was calculated over the entire test set using the batch weight vector $\mathbf{w}_m$.

4.3.2 On-Chip Learning Behavior

The measured learning curves in Fig. 4.8a show that convergence is faster at a higher learning rate γ for both $N = 32$ and $N = 64$. In addition, the learning curves become smoother and converge to a lower p_e with larger batch sizes, e.g., for $\gamma = 2^{-4}$, the misclassification rate p_e at $m = 400$ is lower for the batch size $N = 64$ than for $N = 32$. Figure 4.8b shows that the fixed-point algorithm converges for a regularization factor $\lambda = 2^{-4}$ but diverges for values of $\lambda = 2^{-1}$ and $\lambda = 2^{-5}$. For $\lambda = 2^{-5}$, the weights initially converge and then diverge due to overflow, where as for $\lambda = 2^{-1}$, the algorithm does not converge at all, because the MSBs that are used in the SVM dot product (4.1) remain at zero.

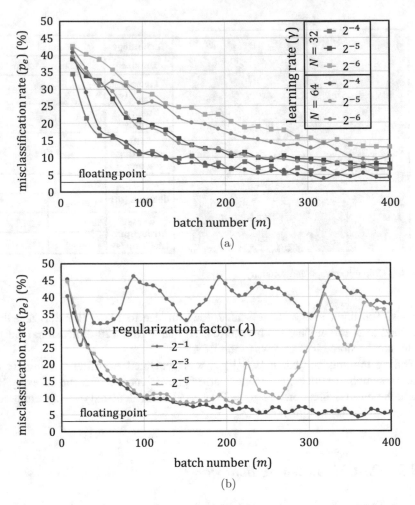

Fig. 4.8 Measured learning curves showing the impact of: (**a**) the batch size N and learning rate γ with $\lambda = 2^{-4}$, and (**b**) the regularization factor λ with $\gamma = 2^{-4}$ and $N = 64$

4.3.3   Robustness

Figure 4.9 shows that on-chip learning converges with randomly set initial weights in the BCA. Furthermore, the learning curves converge to within 1% of floating-point accuracy with 400 batch updates for $\gamma = 2^{-3}$ and 2^{-4} with $\Delta V_{\mathrm{BL,max}} = 560\,\mathrm{mV}$. The misclassification rate p_e increases dramatically to 18% when $\Delta V_{\mathrm{BL,max}}$ is reduced to $320\,\mathrm{mV}$ at $m = 400$. As discussed in Sect. 4.2.1.1, that increase is due to the increased impact of spatial variations. Continued on-chip learning reduces p_e down to 8% for $\gamma = 2^{-4}$ and 2^{-3} within 150 additional batch updates. Similar results are observed when the illumination of the input

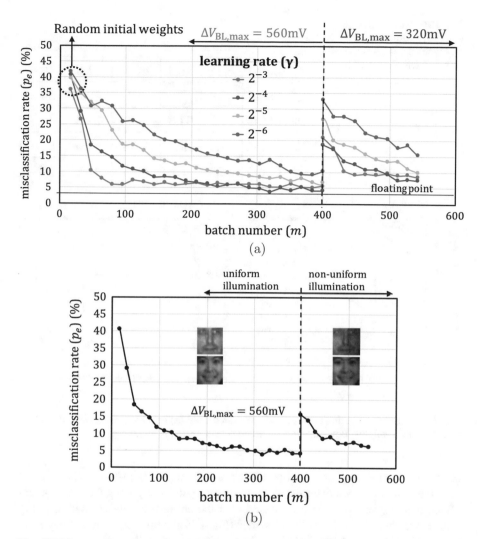

Fig. 4.9 Measured learning curves showing robustness to: (**a**) process variations, and (**b**) variations in input statistics with a batch size $N = 64$ and a regularization factor $\lambda = 2^{-4}$

images changes abruptly at $m = 400$ (see Fig. 4.9b), where p_e increases to 16% and eventually reduces to 6% with further training. The measurements in Fig. 4.9 indicate the effectiveness of on-chip learning in enhancing robustness to variations in the process parameters and the input data statistics.

The chip-specific nature of the learned weights can be seen in Fig. 4.10 which shows that the average p_e increases from 8.4% to 43% when weights learned on a different die are used. This result further highlights the need for on-chip learning.

The receiver operating curve (ROC) characterizes the true positive rate p_{tp} (the fraction of positive examples (faces) classified correctly) with respect to

Fig. 4.10 Misclassification rate (p_e) measured across dies when the weights trained on a different die are used

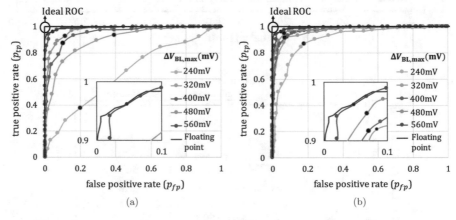

	tested on				
	Chip1	Chip2	Chip3	Chip4	Chip5
Chip1	8.25	38.3	48.3	51.5	48.8
Chip2	45.8	9	48	49.8	34.5
Chip3	47	51.3	8.5	29.8	49.3
Chip4	51.5	51	17.5	8.25	51.3
Chip5	38.3	18	48.5	48.5	8

trained on

Fig. 4.11 Receiver operating curves (ROC) measured with: (**a**) off-chip trained weights, and (**b**) on-chip trained weights. Black markers represent default operating points

false positive rate p_{fp} (the fraction of negative examples (non-faces) classified as positive). The false positive rate p_{fp} was set by digitally adjusting the ADC output offset, and p_{tp} and p_{fp} were measured over the entire test dataset. With off-chip trained weights, the ROC of the classifier (Fig. 4.11a) degrades (moves away from the ideal) with decreasing $\Delta V_{BL,max}$. In addition, the default operating point, i.e., without ADC offset cancellation, (black markers in Fig. 4.11a), are away from the optimal (knee of the ROC). However, with on-chip training (see Fig. 4.11b), the ROCs shift towards the ideal (upper left corner), and the default operating point (black markers) also moves automatically to the optimal location (the knee), thereby eliminating the need for ADC offset tuning.

4.3.4 Energy Consumption

The minimum $\Delta V_{BL,max}$ required to achieve a misclassification rate $p_e \leq 8\%$ without compensating for process variations is 520 mV (see Fig. 4.12a). On-chip

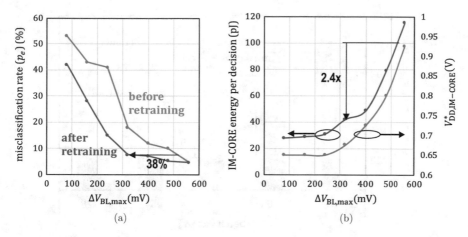

Fig. 4.12 Measured misclassification rate p_e showing 38% reduction in BL swing attributed to on-chip learning, leading to a 2.4× reduction in energy ($V^*_{DD,\text{IM-CORE}}$ is the minimum IM-CORE supply to avoid destructive reads)

training enables the IC to achieve a misclassification rate below 8% at a 38% lower $\Delta V_{\text{BL,max}} = 320\,\text{mV}$. Operating with $\Delta V_{\text{BL,max}} = 320\,\text{mV}$ also enables the IC to operate with a lower $V_{DD,\text{IM-CORE}} = 0.675\,\text{V}$ without destructive reads relative to operating at $V_{DD,\text{IM-CORE}} = 0.875\,\text{V}$ when $\Delta V_{\text{BL,max}} = 520\,\text{mV}$ (see Fig. 4.12b). Thus, on-chip learning enables a reduction in IM-CORE energy by 2.4× at iso-accuracy. That energy gain over the multi-functional DIMA IC described in Chap. 3 ranges from 1.5× to 2.6× for p_e in the range 5% to 8%.

The energy cost of training is dominated by SRAM writes of the updated weights in (4.11) at the end of each batch (see Fig. 4.13). This cost reduces with increasing batch size N, reaching 26% of the total energy cost, for a $N = 128$. At $N = 128$, 60% of the total energy can be attributed to CTRL, whose energy reduces with increasing BCA size.

Figure 4.14 shows the energy breakdown of the prototype IC compared to a conventional digital reference architecture (CONV). CONV is a 2-stage pipelined architecture comprising an SRAM of the same size as in the prototype IC, and a synthesized digital block. The conventional SRAM has a column multiplexer ratio of 4:1, therefore requiring 16× more read accesses than the DIMA. The energy and delay numbers of CONV were based on measured read energy from the prototype IC and computational energy from post-layout simulations. The energy and delay benefits of the prototype IC stem from: (a) simultaneously reading multiple rows and processing them in low swing analog domain; (b) eliminating the 4:1 column mux; and (c) by aggressively reducing BL swing enabled by the use of chip-specific weights obtained via on-chip learning. When operating with pre-trained weights at $\Delta V_{\text{BL}} = 560\,\text{mV}$, the prototype IC shows a 7.8× reduction in energy compared to CONV, where the contributions due to (a) and (b) are estimated to be 5× and 1.6×, respectively. Use of chip specific weights via on-chip learning increases

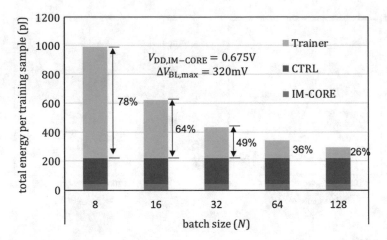

Fig. 4.13 Energy overhead of on-chip training with respect to batch size

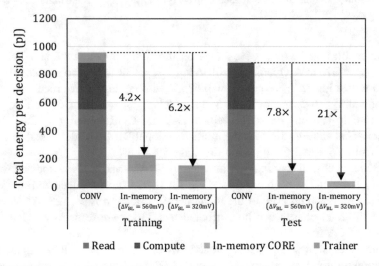

Fig. 4.14 Measured on-chip energy for training (at $N = 64$) and inference compared to a conventional digital reference architecture (CONV), showing $21\times$ reduction in energy consumption with a simultaneous $4.7\times$ reduction in delay, leading to a $100\times$ reduction in EDP during inference. The supply $V_{DD,IM\text{-}CORE}$ is 1 V and 675 mV when operating with a ΔV_{BL} of 560 and 320 mV, respectively

the energy reduction to $21\times$. Along with the reduction in energy, the prototype IC simultaneously shows an overall $4.7\times$ reduction in delay, thereby achieving an overall $100\times$ reduction in EDP during inference. Due to the use of digital read and write operations during the weight update, the energy gain during training reduces to $6.2\times$ (from $21\times$ during inference) at a batch size $N = 64$.

Table 4.1 compares the prototype IC to other in-memory dot product implementations. The prototype IC operates with an energy efficiency of 42 pJ/decision at

Table 4.1 Comparison with prior in-memory dot product implementations

	[31]	[32]	[34]	[36]	This work
Technology	130 nm	65 nm	65 nm	65 nm	65 nm
Algorithm	AdaBoost	SVM	CNN	DNN	SVM
Dataset	MNIST	MIT-CBCL	MNIST	MNIST	MIT-CBCL
On-chip memory (kB)	2	16	2	0.5	16
Bitcell type	6T	6T	10T	6T[a]	6T
Energy/decision (pJ)	600	400	–	–	42
Decisions/s	7.9M	9.2M	–	–	32M
Precision[b] ($B_X \times B_W$)	5×1^s	8×8	7×1^s	1×1	8×8^s
Efficiency[c] (TOPS/W)	350	1.25	14	55.8	3.125 (1.07)[d]
Throughput[c] (GOPS)	819	4.17	5.35	1780	4.13
E_{MAC}[e] (pJ)	0.003	0.8	0.071	0.018	0.32 (0.92)[d]
$E_{MAC,p}$[f] (fJ)	0.56	12.5	10.2	17.9	4.9 (14.5)[d]

[a]Separate left and right word lines (WLs) used
[b]Signed number indicated by ^{s}
[c]1 OP = 1 Multiply and accumulate
[d]At $\Delta V_{BL} = 320\,\text{mV} (560\,\text{mV})$
[e]E_{MAC} is the energy of a single MAC operation
[f]Precision-scaled MAC energy $E_{MAC,p} = E_{MAC}/(B_X \times B_W)$

a throughput of 32 M decisions/s, which corresponds to a computational energy efficiency of 3.12 TOPS/W (1 OP = one 8-b×8-b MAC) for inference. DIMA's energy-latency benefits arise primarily from reduced memory access costs, which tend to dominate with large SRAM bank sizes. Furthermore, DIMA is best suited for algorithms that that suffer most from memory access costs such as fully connected deep neural networks (FC-DNN). Therefore, the prototype IC is compared with [7] which implements a FC-DNN, and employs aggressive voltage/frequency scaling along with digital error compensation techniques such as RAZOR [17]. Employing reported arithmetic efficiency of 0.09–0.16 TOPS/W [7] and accounting for the difference in the process node (28 nm FD-SOI [7] vs. 65 nm bulk CMOS), it is observed that the prototype IC achieves a 30× savings in energy accompanied by a 1.8× savings in delay to implement an 8-b 128-wide dot product. Those energy savings demonstrate the suitability of the proposed architecture for energy-constrained sensory IoT applications.

4.4 Conclusion

This chapter has presented an IC realization of a deep in-memory classifier for the SVM algorithm with on-chip learning in a 65 nm CMOS process. On-chip training overcomes the impact of variations in both process parameters and data statistics, thereby enabling the IC to operate at lower BL discharges than otherwise possible thereby saving energy.

This chapter, demonstrates that on-chip learning is effective in compensating for process variations in DIMA. However, to take full advantage of this technique, a number of algorithmic, architectural, and circuit challenges need to be overcome. Unsupervised and semi-supervised learning algorithms are needed to avoid having to store the training set on-chip. Efficient write-back techniques would be required in order to minimize the energy consumption during training especially for always-on systems that need to continuously track their environment. Realizing on-chip training is made much more challenging for large-scale deep networks. However, since both inference and trainer computations are based on matrix-vector operations, there is significant potential for employing DIMA to realize both the forward and backward parts of the network. In such cases, the impact of PVT variations on the learning behavior and inference accuracy will be interesting to study.

Chapter 5
Mapping Inference Algorithms to DIMA

Chapter 3 showed that algorithms with simple and similar algorithmic data-flows can easily be mapped to DIMA. This chapter shows that algorithms with very different and significantly complex data-flow can also be mapped to DIMA. In particular, mapping of the following two algorithms to DIMA is shown: (1) a convolutional neural network (CNN) [41], and (2) a sparse distributed memory (SDM) [43, 74] where the latter is a computational model inspired by the human brain. Opportunities to compensate for DIMA's non-ideal analog computations algorithmically in exist both enabling further reduction in the compute SNR (low-SNR operation) to achieve aggressive energy savings, (e.g., the use of error-aware training in DIMA-based CNN) and the use of ensemble decision-making in SDM.

5.1 Convolutional Neural Network (CNN)

Convolutional neural networks (CNN) [75] are a commonly used inference algorithm for computer vision tasks, such as handwriting recognition and face detection [76, 77] because of its state-of-the-art performance. However, CNNs tend to be very complex in terms of their parameter size and number of computations, e.g., requiring the input data to be processed by more than 100 million weights [78] in order to generate a single M-ary decision. Thus, the implementation of CNNs requires large the computational and storage requirements. GPU [76] and FPGA-based [77] implementations have been proposed to speed up CNN computation. It is also well known that the energy efficiency and throughput of general-purpose/reconfigurable computing platforms such as GPUs and FPGAs, are at least one to two orders of magnitude worse than those of accelerators [29]. DIMA-based CNN implementation in [41] has been shown to provide a 24.5× reduced EDP as compared to the conventional system.

© Springer Nature Switzerland AG 2020
M. Kang et al., *Deep In-memory Architectures for Machine Learning*,
https://doi.org/10.1007/978-3-030-35971-3_5

5.1.1 CNN Overview

A CNN is a multi-layer network (see Fig. 5.1a) consisting of a cascade of many convolutional layers (C-layers) followed a few fully connected layers (FC-layers). The C-layers tend to be compute-bound, since the weight parameters are shared across multiple outputs, while the FC-layers are memory bound, since FCs implements a full MVM in which each output activation has its weight parameter vector. A C-layer with M input channels (or feature maps (FMs)) and N output channels is described as follows:

$$\begin{bmatrix} \mathbf{y}_1 \\ \vdots \\ \mathbf{y}_N \end{bmatrix} = \phi \left\{ \begin{bmatrix} \mathbf{w}_{11} & \cdots & \mathbf{w}_{1M} \\ \vdots & \cdots & \vdots \\ \mathbf{w}_{N1} & \cdots & \mathbf{w}_{nm} \end{bmatrix} * \begin{bmatrix} \mathbf{x}_1 \\ \vdots \\ \mathbf{x}_m \end{bmatrix} + \begin{bmatrix} C_1 \\ \vdots \\ C_N \end{bmatrix} \right\} \tag{5.1}$$

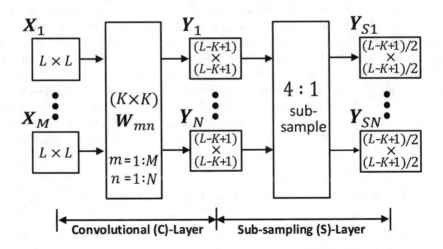

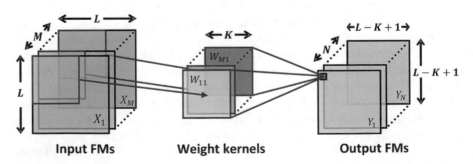

Fig. 5.1 Convolutional neural network (CNN) structure: (**a**) data flow, and (**b**) the convolutional (C) layer

where $\mathbf{x}_m$ ($m = 1, \ldots, M$) and $\mathbf{y}_n$ ($n = 1, \ldots, N$) are the $L \times L$ input FMs and $(L - K + 1) \times (L - K + 1)$ are the output FMs, respectively; $*$ is the convolution operator; and $\mathbf{w}_{nm}$ is a $K \times K$ weight kernel that contributes to the generation of $\mathbf{y}_n$ (the nth output FM) via convolution with $\mathbf{x}_m$ (the mth input FM). Here, C_n is a bias term for the nth output FM, and ϕ is a non-linear kernel, typically sigmoid, tanh, or a ReLU function. There is usually a pooling or subsampling layer (S-layer) that reduces the FM dimensions via either averaging of neighboring activations (average pooling) or choosing of their maximum (max pooling). Typical values for a state-of-the-art CNN ResNet [79] is $M = 64$–2048, $K = 3$, and $L = 7$–112 with the number of layers up to 1000.

Figure 5.1b shows the convolutional layer, in which $\mathbf{y}_n$ is generated by sliding the $K \times K$ input window (red box) across $\mathbf{x}_m$ and summing the output of M such channels. Here, $Y_n(p, q)$ is the (p,q)-th element of $\mathbf{y}_n$, computed as follows:

$$\widetilde{Y}_n(p, q) = \sum_{m=1}^{M} \sum_{i=1}^{K} \sum_{j=1}^{K} W_{nm}(i, j) X_m(p + i, q + j)$$

$$Y_n(p, q) = \phi \left\{ \widetilde{Y}_n(p, q) + C_n \right\} \tag{5.2}$$

where $W_{nm}(i, j)$ and $X_m(i, j)$ are the (i, j)th elements of $\mathbf{w}_{nm}$ and $\mathbf{x}_m$, respectively. Since the C-layer requires a large number of read accesses and computations, the proposed DIMA-CNN is highly effective in realizing CNNs.

5.1.2 CNN Implementation Challenges

A number of challenges need to be addressed to enable energy-efficient and high-throughput realization of CNNs. In the following, VGGNet [78] parameters are employed to illustrate these challenges.

(A1) Frequent Data Access and Processing: Equation (5.2) shows that large data volumes need to be read from memory and processed, e.g., MNK^2 (up to 3×10^5 [78]) bytes of data access and $MNK^2(L - K + 1)^2$ (up to 2×10^9 [78] multiply-and-accumulation (MAC)) operations per C-layer. Therefore, approaches for energy-efficient data access and processing are required.

(A2) Realizing Sliding Window Functionality for Convolutions: To realize the sliding window function shown in Fig. 5.1b, efficient data access is required to provide the appropriate window input FMs $\mathbf{x}_m$ to the processor for convolution ($*$) operations.

(A3) Enabling Data Reuse in Analog: As indicated in Fig. 5.1b, the weight kernel $\mathbf{w}_{nm}$ can be reused up to $(L - K + 1)^2$ times across the window sliding once it has been fetched from memory. Analog-heavy architectures make data reuse challenging as analog computations are vulnerable to various noise sources, such as leakage and coupling noise.

(A4) Parallelizing Convolutions: As the C-layer requires a large number of MAC operations (up to 2×10^9 [78] per decision), massive parallelism is required to achieve high throughput. It is challenging to develop an efficient data storage format and architecture to expose and exploit parallelism.

5.2   Mapping CNN on DIMA (DIMA-CNN)

A DIMA-based CNN (DIMA-CNN) addresses the implementation challenges described in Sect. 5.1.2 as follows

- DIMA with its high energy efficiency and intrinsic parallelism addresses **(A1)** and **(A4)**,
- a sliding window FM register and optimized FM access patterns addresses **(A2)**,
- a charge-recycling mixed-signal multiplier enabling the reuse of functional READ outputs addresses **(A3)**,
- multiple bitcell array banks and an optimized kernel data storage format to address **(A4)**.

The architecture and operation of DIMA-CNN are described next.

5.2.1   A Multi-Bank DIMA-CNN Implementation

Figure 5.2a shows a single-bank DIMA-CNN comprising the bitcell array, BL processing (BLP units), cross BL processing (CBLP), ADC, and input first-in, first-out (FIFO) registers. The $N_{\text{ROW}} \times N_{\text{COL}}$ bitcell array is partitioned into $N_{\text{sub}} = \lfloor N_{\text{COL}}/(2K^2) \rfloor$ sub-arrays, each of which has $2K^2$ columns (the factor 2 is from the use of sub-ranged functional READ), to store $\mathbf{w}_{nm}$ whereas the input FIFO stores the K^2 pixels of the current input window in $\mathbf{x}_m$ (the red box in Fig. 5.1b).

Figure 5.2b describes the multi-bank DIMA-CNN architecture including N_{bank} banks, the RDL block, and the FM register bank. The output FMs of the S-layers are stored in the FM register to avoid the high energy and delay cost of the write operations into the SRAM bitcell array. The FM register also supports the sliding window functionality, which is difficult to realize in a SRAM. The FM register bank includes an input register (I-REG) to store the input image pixels, and N_1 FM1 registers (FM1-REG) and N_2 FM2 registers (FM2-REG) to store the output FMs of the first and second S-layers, respectively. Here, N_1 and N_2 are the number of output FMs of the first and second S-layers, respectively. The output FM of subsequent layers, e.g., the fully-connected layer, can be stored in the FM1-REG through overwriting of its contents, as the previous layer's output FMs are not required once the output FM of the current layer has been computed.

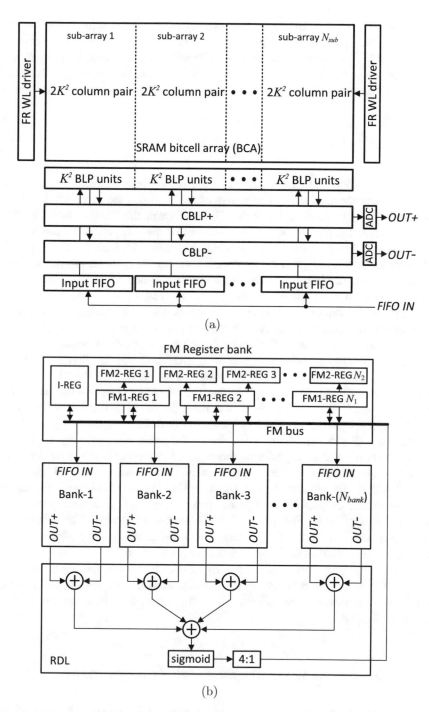

Fig. 5.2 DIMA-based architecture for CNN with: (**a**) single bank, and (**b**) multiple (N_{bank}) banks, where RDL comprises an accumulator, a sigmoid, and 4:1 sub-sampling blocks

DIMA's functional READ operation is inherently 1-dimensional (1-D), whereas the convolution operation in (5.2) is a 2-dimensional (2-D) operation. Therefore, the convolution in (5.2) is reformulated as a 1-D operation by transforming the $K \times K$ 2-D kernel $\mathbf{w}_{nm}$ and the $K \times K$ input window in $\mathbf{x}_m$ (starting at the (p, q)th pixel) into K^2-dimensional 1-D vectors $\mathbf{w}_{1D,nm}$ and $\mathbf{x}_{1D,m}(p, q)$, respectively, as follows:

$$\tilde{Y}_{1D,n}(p, q) = \sum_{m=1}^{M} \sum_{i=1}^{K^2} W_{1D,nm}(i) X_{1D,m}(p, q, i) \tag{5.3}$$

where $W_{1D,nm}(i)$ and $X_{1D,m}(p, q, i)$ are the ith element of vectors $\mathbf{w}_{1D,nm}$ and $\mathbf{x}_{1D,m}(p, q)$, respectively. The FM register bank provides the input window in $\mathbf{x}_m$ starting at (p, q)th pixel to the input FIFOs through the FM bus and the *FIFO IN* port so that the FIFO contains the 1-D vector $\mathbf{x}_{1D,m}(p, q)$. Each sub-array in Fig. 5.2a processes the inner summation of (5.3), while N_{sub} such sub-arrays process the outer summation of (5.3) by aggregating the results from all the sub-arrays in the cross BL processing block via charge-sharing. When $M > N_{sub}$, the outputs of ADCs in each bank are accumulated in RDL (Fig. 5.2b) in the digital domain to avoid the degradation in the analog outputs during its transmission through lossy inter-bank interconnects. An appropriate data storage pattern and processing sequence will be presented in Sect. 5.2.2 for the case when $M \neq N_{sub}$.

Each BL processing unit multiplies $W_{1D,nm}(i)$ from the functional READ and $X_{1D,m}(p, q, i)$ from the input FIFO, followed by aggregation of the BL processing results in the cross BL processing blocks. The BL processing outputs are charge-shared on either CBLP+ or CBLP- rails in Fig. 5.2a, based on the sign of the weight detected by the comparator, which is shown in Fig. 5.3.

The RDL (Fig. 5.2b) accumulates the partial sums computed by N_{bank} such banks if needed, and then performs the sigmoid function, which is implemented digitally using a piecewise linear approximation composed of three additions and two shifts [77]. To perform the 4:1 sub-sampling operation S-layer computation, four consecutive outputs of the sigmoid processing block, that correspond to the four neighboring output FM pixels of the C-layer, are averaged. Those outputs of the S-layer computation are fed back into the corresponding FM registers through the FM bus to be used as the input FM of the next C-layer.

Note that the proposed architecture realizes the most compute-intensive operation in a CNN, the dot-products in (5.3), in the analog domain to achieve energy and delay efficiency, and realizes the rest in the digital domain in order to provide maximum flexibility.

5.2.2 Efficient Data Storage for Parallel Computations

The data storage format shown in Fig. 5.4 aims to compute (5.3) via a single functional READ and BL processing step. The K^2 coefficients of $\mathbf{w}_{1D,nm}$ are stored

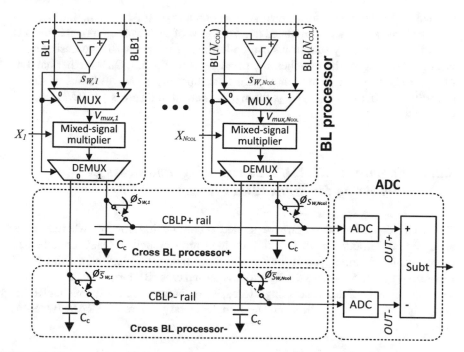

Fig. 5.3 BL processing and cross BL processing architecture for signed dot product

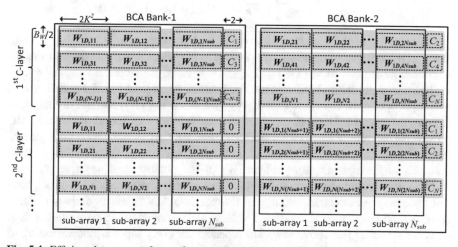

Fig. 5.4 Efficient data storage format for a two-bank ($N_{bank} = 2$) configuration

horizontally in a block of $(B_W/2) \times 2K^2$ bitcells (shown as dotted boxes in Fig. 5.4), where a B_W-bit word is stored per two columns (column pair) to enable sub-ranged reads (Sect. 2.4.1). Accordingly, a total of $N_{bank}N_{sub}$ sub-arrays can be processed per read access. The $\mathbf{w}_{1D,nm}$s with identical values of index n are horizontally aligned in the same word-row occupying M sub-arrays. If $M < N_{sub}$, the $\mathbf{w}_{1D,nm}$s with the same N are accommodated in a word-row of each bank as shown in the 1-st row of Fig. 5.4. Empty sub-arrays can be left out of the cross BL processing stage through disabling of the switches ϕ_{S_W}s and $\phi_{\overline{S}_W}$s in Fig. 5.3. On the other hand, if $M > N_{sub}$, the $\mathbf{w}_{1D,nm}$s are stored across multiple (up to N_{bank}) banks as shown in fourth row. If $M > N_{bank}N_{sub}$, the $\mathbf{w}_{1D,nm}$s are stored across $\lceil M/(N_{bank}N_{sub})\rceil$ word-rows.

5.2.3 Functional READ, BL Processing, and Cross BL Processing for Signed Dot Product

The CNN computation requires a signed representation for W, whereas X is an unsigned number (m, n, i, and j omitted for simplicity). One's complement representation is employed to obtain the magnitude ($|W|$) and sign (S_W) of W separately. Figure 5.3 shows the BL processing and cross BL processing architecture to enable signed number processing. Here, $|W|$ is computed in the functional READ stage through the exploitation of the complementary nature of SRAM bitcell as follows:

$$|W| = \begin{cases} \Sigma_{k=0}^{B_W-1} 2^k w_k \propto \Delta V_{BLB}(W), & \text{if } W \geq 0 \\ \Sigma_{k=0}^{B_W-1} 2^k \overline{w}_k \propto \Delta V_{BL}(W), & \text{if } W < 0 \end{cases} \qquad (5.4)$$

The sign S_W is obtained by a differential analog comparator with $\Delta V_{BL}(W)$ and $\Delta V_{BLB}(W)$ as its inputs. The sign S_W is used as the select signal into the multiplexer to choose the greater of $V_{BL}(W)$ and $V_{BLB}(W)$. Thus, the output $\Delta V_{mux} \propto |W|$.

The mixed-signal multiplier in the BL processing stage of Fig. 5.3 operates with two input operands: the analog input V_{mux} (representing $|W|$) and the digital input X from the input FIFO. Next, the multiplier outputs are transferred on the cross BL processing capacitors C_c of either the CBLP+ or the CBLP- rail via de-multiplexers controlled by the comparator sign output S_W. The rails are shared with multiple columns so that the absolute values $|W|$ of the positive and negative products in (5.3) can be added separately. Finally, the charge-shared cross BL processing rail voltages are converted into digital through two ADCs, which outputs positive (*OUT+*) and negative (*OUT-*) digital values, which are added to generate the final convolution sum. The above steps are repeated for every input window, leading to significant energy consumption, especially from the functional READ process. To avoid the functional READ iterations, a charge-recycling mixed-signal multiplier is introduced as described in the following section.

5.2.4 Charge-Recycling Mixed-Signal Multiplier

Figure 5.5 shows the charge-recycling mixed-signal multiplier that computes the product of analog input V_{in} and a $B_X(= 3)$-bit digital value X. In spite of being implemented in analog, this multiplier is able to exploit the intrinsic data reuse opportunities provided by a CNN's data-flow, as described below.

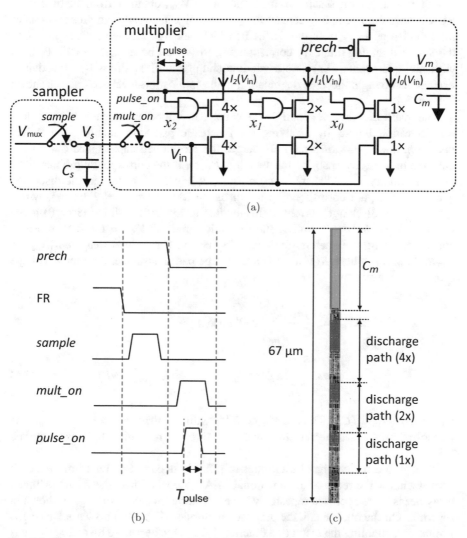

(a)

(b)   (c)

Fig. 5.5 Charge-recycling mixed-signal multiplier to enable data reuse with bit-precision $B_X = 3$: (a) multiplier circuit diagram with a sampler, (b) timing diagram, and (c) layout fabricated in the 65 nm CMOS prototype IC of [32] as a separate test module

A single unit of the mixed-signal multiplier was fabricated in the prototype IC of [32] as a separate test module to investigate its potential use for convolution operation. The multiplier is designed to be pitch-matched to the horizontal dimension of a bitcell (Fig. 5.5c) to enable column-wise parallelism. In practice, $B_X = 6$ is employed in this section by cascading two (3-b MSB and 3-b LSB) multipliers in parallel to reduce the delay. Outputs of these multipliers are merged in the cross BL processing stage, and are combined with $8\times$ higher weight given to the MSB outputs [44]. The multiplier first samples the analog input V_{mux} obtained from the functional READ stage on a sampling capacitor C_S (Fig. 5.5b). The sampling process isolates the sampled node V_s from the BL/BLB, which has high leakage paths. Therefore, the sampled analog value V_s corresponding to W can be reused up to 200 times with multiple inputs (Xs) across window slidings until V_s drops by 10% due to leakage incurring noticeable degradation in the application-level accuracy as shown in Sect. 5.4.

The multiplier has B_X pull-down paths corresponding to each bit position. Multiplication begins by disabling the precharge path and enabling the switch *mult_on*. Each pull-down path discharges the precharged capacitance C_m for the duration of T_{pulse} by enabling the switch *pulse_on* if the corresponding binary data $x_i = 1$. Binary weighted transistor sizing (see Fig. 5.5a) results in the discharge current of the path corresponding to ith bit position $I_i(V_{in}) = 2^i I_0(V_{in})$, where $I_0(V_{in})$ is the discharge current corresponding to the LSB position ($i = 0$) given by $I_0(V_{in}) = bV_{in} + c$ within the dynamic range of $V_{in} = 0.6$–1 V, where b and c are fitting parameters, (analyzed in Sect. 5.3). Provided that the $T_{pulse} \ll C_m R_m(V_{in}, X)$, where $R(V_{in}, X)$ is a discharge path resistance, the output voltage drop ΔV_m is given by:

$$\Delta V_m = \frac{T_{pulse}}{C_m} \sum_{i=0}^{B_X-1} x_i 2^i I_0(V_{in})$$

$$= a \sum_{i=0}^{B_X-1} x_i 2^i (bV_{in} + c) = abX(V_{in} + c/b) \tag{5.5}$$

where $a = T_{pulse}/C_m$. Thus, $\Delta V_m \propto XV_{in}$ with an offset c/b added to V_{in}. The impact of offset is analyzed and overcome through retraining [80] as described Sect. 5.4.

Though the mixed-signal multiplier in [32, 41] does not suffer from offsets, it does not allow the reuse of the functional READ outputs. Thus, the W in the bitcell array needs to be accessed repeatedly for every kernel window sliding, as shown in Fig. 5.6a. On the other hand, the proposed multiplier (Fig. 5.5a) does not have such limitations, enabling the result of a functional READ to be reused up to 200 times as shown in Fig. 5.6b, thereby minimizing the number of functional READ operations.

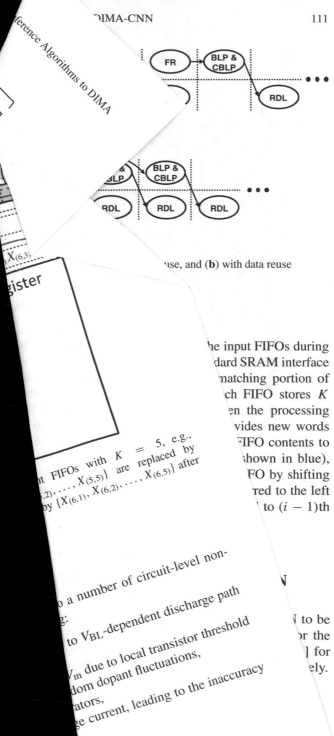

$X_{(6,3)}$

gister

ise, and (**b**) with data reuse

he input FIFOs during
dard SRAM interface
matching portion of
ch FIFO stores K
en the processing
vides new words
FIFO contents to
shown in blue),
FO by shifting
red to the left
to $(i-1)$th

t FIFOs with $K = 5$, e.g.,
,2), $\ldots, X_{(5,5)}$ are replaced by
by $\{X_{(6,1)}, X_{(6,2)}, \ldots, X_{(6,5)}\}$ after

o a number of circuit-level non-
g:
to V_{BL}-dependent discharge path
V_m due to local transistor threshold
dom dopant fluctuations,
ators,
ge current, leading to the inaccuracy

N

N to be
or the
] for
ely.

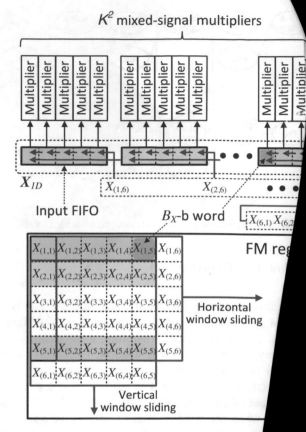

Fig. 5.7 FM register to enable window sliding and inpu
the green-marked right most FIFO contents $\{X_{(5,1)}, X_{(5}$
$\{X_{(5,2)}, X_{(5,3)}, \ldots X_{(5,6)}\}$ after horizontal window sliding or
vertical sliding

5.3.1   Functional Model

DIMA is analog-intensive and therefore subject t
idealities. Dominant among them are the followin

- non-linearity of functional READ process due
  resistance R_{BL},
- impact of process variations on ΔV_{BL} and Δ
  voltage V_{t}-mismatch caused primarily by ran
- input offset of the analog differential compa
- charge loss on the node V_{s} due to the leaka
  on the node V_{in},

- mixed-signal multiplier offset (c/b in (5.5)) due to the non-ideal transistor switching behavior.

The non-linearity of functional READ in $B1)$ is modeled by a polynomial fit of measured results in [32, 44] as $\Delta \widetilde{V}_{BL}(W) = \sum_{k=0}^{5} c_k W^k$, where $\Delta \widetilde{V}_{BL}(W)$ is a distorted version of $\Delta V_{BL}(W)$, and c_ks are the fitting parameters. The impact of process variations in $B2)$ is modeled as a Gaussian distributed random variable (RV) as shown below:

$$\Delta \widehat{V}_{BL}(W) \sim \mathcal{N}(\Delta \widetilde{V}_{BL}(W), \sigma_{BL}^2(W)) \tag{5.6}$$

where $\sigma_{BL}^2(W)$ is the variance of $\Delta \widehat{V}_{BL}(W)$. The same modeling method is applied for the analog multiplier with $\sigma_m^2(V_{in})$, the variance of ΔV_m in (5.5). Monte Carlo SPICE simulations show that $\sigma_{BL}(W)$ and $\sigma_m(V_{in})$ are less than 12.5 and 6.5% of their average values, respectively. Note that the noise due to the variation is well averaged out during the aggregation of 128 BL processing outputs in the cross BL processing stage. The measurement results of prototype IC in [32] confirms this as the standard deviation of cross BL processing output across 128 rows is less than 1.1% of the average value.

The comparator offset in $(B3)$ is modeled as:

$$V_{mux}(W) \sim \max(V_{BL}(W), V_{BLB}(W) + \eta) \tag{5.7}$$

where η is a zero-mean Gaussian distributed RV with 10 mV standard deviation [29].

The impact of leakage in $(B4)$ on the node V_{in} as a function of the reuse number r is modeled as follows:

$$\widehat{V}_{in} = V_{in}e^{-\gamma(V_{in})r} \simeq V_{in}(1 - \gamma(V_{in})r) \tag{5.8}$$

where $\gamma(V_{in})$ is the discharge rate of V_{in} after each reuse. Provided that $r \ll 1/\gamma(V_{in})$, the impact of leakage can be approximated as shown in (5.8). SPICE simulations show that $\gamma < 0.05\%$ at the FF corner and at 85 °C, where leakage current is maximized. This also includes the kick-back noise from the toggling of switch $mult\_on$ to the node V_s. The parameter r is uniformly distributed, i.e., $r \sim U(1, R)$, where $R \ll 1/\gamma(V_{in})$ is the maximum allowed reuse number.

Figure 5.8 shows the measured results for the mixed-signal multiplier fabricated in a 65 nm process that is used to model the non-ideality $(B5)$. The model parameters in (5.5) can be written as $\alpha = ab$, and $\beta = c/b$, and can be fitted to the curve in Fig. 5.8 with $\alpha = 0.16$, and $\beta = -0.5$. The maximum modeling error is less than 2% of ΔV_m's dynamic range. The offset β results in a deviation from the ideal product $\alpha X V_{in}$, but since it is deterministic, it can be compensated via offline retraining of $\mathbf{w}_{nm}$ and C_n in (5.1) can compensate for it.

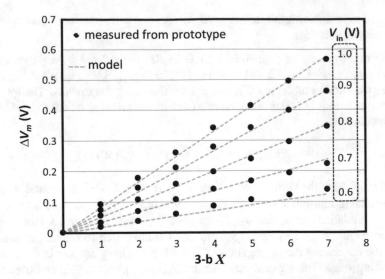

Fig. 5.8 Measured accuracy of the mixed-signal multiplier fabricated in the prototype IC of [32] compared with the model in (5.5) with $\alpha = ab = 0.16$ and $\beta = c/b = -0.5$

5.3.2   Delay and Energy Models

This section provides the delay and energy models for the conventional and proposed DIMA architectures. We focus on the C-layer, but these models can also be applied to the F-layers with slight modification. The conventional system generates N output FMs with a delay given by:

$$T_{\mathrm{conv}} \approx \left\lceil \frac{MNK^2}{(B_{\mathrm{IO}}/B_W)N_{\mathrm{bank}}} \right\rceil T_{\mathrm{read}} + \left\lceil MNK^2/N_{\mathrm{mult}} \right\rceil N_{\mathrm{mov}}T_{\mathrm{mult}} \qquad (5.9)$$

where $N_{\mathrm{mov}} = (L - K + 1)^2$ is the number of window slidings across input FM, B_{IO} is the bit width of SRAM IO so that B_{IO}/B_W (e.g., 2–8) words are fetched per read access per bank, N_{mult} is the number of digital multipliers, and T_{read} and T_{mult} are the delays per the memory access and multiplication, respectively. The delay and energy for the aggregation via an adder tree is assumed to be negligible. The first and second terms in (5.9) correspond to the delays of SRAM read and subsequent convolution operations. Similarly, DIMA-CNN generates N output FMs with a delay given by:

$$T_{\mathrm{DIMA}} \approx \left\lceil \frac{MNK^2}{(N_{\mathrm{bank}}N_{\mathrm{COL}}/2)} \right\rceil \times (\lceil N_{\mathrm{mov}}/R \rceil T_{\mathrm{FR}} + N_{\mathrm{mov}}T_{\mathrm{BLP}}) \qquad (5.10)$$

where T_{FR} and T_{BLP} are the delays of functional READ and BL processing stages, respectively. DIMA can access and process $N_{\mathrm{COL}}/2$ (e.g., 128) words per bank in parallel, where factor 2 in the denominator arises from the use of sub-ranged read (see Sect. 2.4.1). This parallelism creates a significant throughput enhancement, as we can see by comparing the first terms in (5.10) and (5.9).

The energy consumption of the conventional architecture in generating N output FMs is given by:

$$E_{conv} \approx MNK^2 E_{read} + MNN_{mov} E_{reg} + MNK^2 N_{mov} E_{mult} + P_{leak} T_{conv} \quad (5.11)$$

where E_{read}, E_{mult}, and P_{leak} represent the single word SRAM read energy, digital multiplier energy, and SRAM leakage power consumption, respectively. It is assumed that a deep-sleep mode is enabled to minimize the leakage during standby through the use of techniques such as power gating or lowering of the supply voltage for the bitcell array [46]. Here, E_{reg} is the register energy to: (1) fetch five words per window sliding from the FM register, (2) shift the contents of input FIFOs per window sliding in Fig. 5.7, and (3) write back a word of the output FM into the FM register after four such window shifting and sub-sampling. Similarly, the energy consumption of DIMA is given by:

$$E_{DIMA} \approx MNK^2 \lceil N_{mov}/R \rceil E_{FR} + MNN_{mov} E_{reg}$$
$$+ MNK^2 N_{mov} E_{BLP} + P_{leak} T_{DIMA} \quad (5.12)$$

where E_{FR} and E_{BLP} are the energy consumed for the functional READ and BL processing operations of a single word, respectively. They are significantly smaller (by roughly $10\times$) than E_{read} and E_{mult} due to the DIMA's read and processing with low-voltage swing. These models are employed in Sect. 5.4 to demonstrate the benefits of the proposed architecture.

The parameter values $T_{FR} = 7$ ns and $T_{BLP} = 17$ ns in (5.9)–(5.12) are validated via the prototype IC in [32] through the measurement functional READ and BL processing accuracies within the given delay. The conventional SRAM read of the prototype IC in [32] takes 9 ns ($= T_{read}$) per single read access. However, it could be improved if a self-timed controller is employed. Thus, for the conventional system, $T_{read} = 4$ ns is assumed to be conservative. The digital blocks in the conventional system are also synthesized to have the same delay ($T_{mult} = 4$ ns) as T_{read} to enable two-stage pipeline between the dot product processing and the SRAM access in F-layers. The values of E_{FR}, E_{read}, E_{BLP}, and P_{leak} are also validated by the comparison of the measured total energy of prototype IC in [32] and that from the post-layout simulations. Other parameters for digital blocks such as E_{mult} and E_{reg} are estimated from post-layout SPICE simulations of synthesized blocks. The delay, energy, and behavioral model parameters are summarized in Table 5.1.

The F-layers are implemented much like the C-layers except that neither the sliding window nor the reuse of functional READ outputs is required. Therefore, the parameters of the F-layer model need to be set accordingly, e.g., $r = 0$ in (5.8), and $N_{mov} = 1$, $R = 1$ in (5.9)–(5.12).

Table 5.1 Delay, energy, and behavioral model parameters

Parameter	Values	Parameter	Values
T_{FR}, T_{read}	7 ns, 4 ns	T_{BLP}	17 ns
T_{pulse}	2 ns	T_{mult}	4 ns
E_{BLP}, E_{mult}	0.08 pJ, 0.9 pJ	E_{FR}, E_{read}	0.5 pJ, 5.2 pJ
E_{reg}, P_{leak}	4 pJ, 2.4 nW	C_S,C_c,C_m	25, 25, 100 fF
$\sigma(W)$	7–12.5% of avg.	$\sigma_m(V_{in})$	2.5–6.5% of avg.
α, β	0.16, −0.5	γ	<0.05%
$c_0, \ldots, c_5$	$-0.04, 0.97, -0.14, 4.7\times10^{-2}$ $-5.3\times10^{-3}, 2.5\times10^{-4}, -4.3\times10^{-6}$		

5.4 Simulation and Results

This section describes the system configuration and simulated results for the proposed DIMA-CNN architecture in terms of its energy, delay, and inference accuracy. As shown in Sect. 5.3.1, the accuracy analysis needs to incorporate the stochastic nature of DIMA, and thus requires a sufficiently large number of simulations. Therefore, we employ LeNet-5 [75] for hand written digit recognition with the MNIST database [64].

5.4.1 Architecture Configurations

LeNet-5 [75] has six layers: C-layers C1 and C3, F-layers F5 and F6, and S-layers S2 and S4. The design parameters of DIMA-CNN are summarized in Table 5.2. Bit precision of $B_W = 8$ and $B_X = 6$ are chosen to obtain negligible accuracy degradation compared to a floating point realization. The DIMA-CNN with 64 KB SRAM ($N_{bank}N_{ROW}N_{COL} = 4 \times 512 \times 256$) and a total of 2 KB registers are employed, where 20 $\mathbf{w}_{mn}$s ($N_{COL}N_{bank}/(2K^2) \approx 20$ with $K = 5$) are stored per word-row and processed per read access cycle. The FM registers for C-layers are designed to support the vertical and horizontal window movements with kernel size $K = 5$. The SRAM, BL processing/cross BL processing blocks, and registers take 79.8%, 8.6%, and 11.6% of overall layout area, respectively.

We assume the conventional architecture comprises a conventional SRAM instead of a functional READ, and digital dot-product computations instead of BL processing and cross BL processing blocks. For the conventional architecture, the IO bit-widths (B_{IO}) per SRAM bank of 16, 32, and 64 bits are considered corresponding to a column mux ratio $L = 4, 8$, and 16, respectively. A $K^2 (= 25)$ dimensional digital dot product block is synthesized and post-layout simulated in the same (65 nm) process. For fair comparison, we assume that the conventional architecture has seven digital dot product blocks as the area occupied by these blocks is approximately equal to the BLP and CBLP blocks in DIMA-CNN. The DIMA-CNN achieves

Table 5.2 Design
parameters of LeNet-5

Parameter	Values	Parameter	Values
V_{DD}	1 V	V_{PRE}	1 V
Input L	32	K	5
B_W, B_X	8, 6	N_{mult}	175
R	0–200	B_{IO}	16–64
$(N)_{bank,row,col}$	4, 512, 256		
N	C1: 6 , C3: 16, F5: 120, F6: 10		

higher parallelism, e.g., the number of BL processing blocks $N_{COL}N_{bank}/2 = 512$, than the conventional architecture with $N_{mult} = 175$ (= 25×7) within the same area due to the pitch-matched BL processing layout as shown in Fig. 5.5c. The energy and delay of the BL processing blocks are estimated via post-layout simulations.

5.4.2 Recognition Accuracy

The error rates (ϵ_{det}) are measured on MNIST test dataset [64] via system simulations for the following architectures:

- FL: floating-point simulations.
- FX: the conventional (SRAM+digital processor) architecture with error-free fixed-point computations (B_W and B_X).
- DIMA: DIMA-CNN with fixed-point computations that use behavioral models (5.5)–(5.8) without resorting to retraining.
- DIMA-TR: DIMA-CNN with fixed-point coefficients retrained accounting for the non-linearity of functional READ process and the offset of charge-recycling mixed-signal multiplier.

Retraining (<10 iterations) employs the behavioral model of the charge-recycling multiplier in (5.5), and is applied after the standard training process (100 iterations). The ϵ_{det}s of DIMA and DIMA-TR are obtained via Monte Carlo simulations with 400 iterations at each value of R. DIMA and DIMA-TR have the same statistical behavior as DIMA-TR is retrained to compensate for only deterministic non-idealities. Figure 5.9 plots the ϵ_{det}s of above architectures for different values of the reuse factors R. The floating-point simulations (FL) and the conventional architecture (FX) achieve a constant $\epsilon_{det} = 0.8$ and 0.97% across R, respectively. In contrast, the ϵ_{det} of DIMA increases with R because of the leakage current as described in (5.8) resulting in a median ϵ_{det} of 1.7% with $R = 800$.

However, the median ϵ_{det} of DIMA-TR reduces by up to 0.2% due to retraining. The ϵ_{det} of DIMA-TR increases negligibly, with a median $\epsilon_{det} < 1.3\%$ (worst-case, $\epsilon_{det} < 2.3\%$) for $R < 200$.

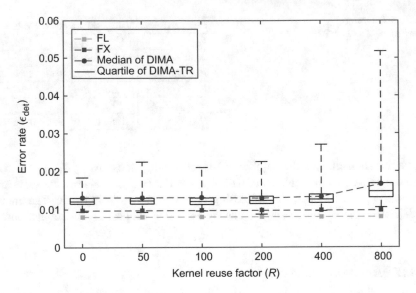

Fig. 5.9 Error rate (ϵ_{det}) vs. reuse factor (R)

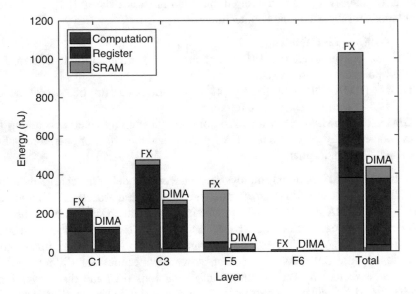

Fig. 5.10 Energy breakdown per layer with $B_{IO} = 16$ and $R = 50$

5.4.3 Energy and Delay

The energy and delay of the conventional architecture and DIMA-CNN are estimated based on (5.9)–(5.12). Figure 5.10 shows the energy breakdowns across the C- and F-layers but not the S-layers, which has a negligible contribution to the total

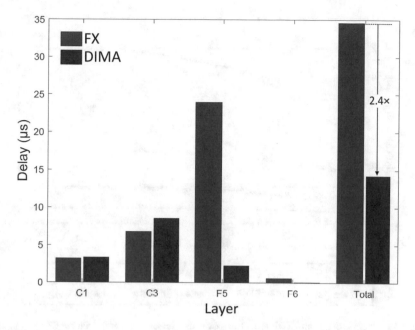

Fig. 5.11 Delay of each layer with $B_{IO} = 16$ and $R = 50$ for the conventional architecture (FX) and DIMA

energy and delay. The DIMA-CNN architecture saves energy in every layer because of the low-swing computation in functional READ, BL processing, and cross BL processing stages. Specifically, DIMA significantly reduces the energy in F5 layer, where the highest data access volume is required compared to the other layers. The overall system energy is reduced by 4.9× when compared to the conventional architecture, achieving an energy efficiency of 436 nJ/decision.

As shown in Fig. 5.11, DIMA-CNN has similar or slightly higher delay in C-layers as compared to the conventional architecture since the proposed mixed-signal multiplier is slower than digital multiplication ($T_{BLP} > T_{mult}$). However, DIMA-CNN achieves a significant throughput improvement in the memory-intensive F5 layer because of the highly parallel functional READ process. Overall, DIMA-CNN achieves approximately a 2.4× delay reduction at a decision latency of 14.3 μs/decision.

Figure 5.12 shows that both energy and delay benefits over the conventional architecture increase with the resue factor R, achieving up to 11.9× EDP reduction. The improvement in EDP saturates at $R = 50$ as the contribution of functional READ to the total energy and delay becomes negligible when $R > 50$. As B_{IO} increases, the energy and delay of the conventional architecture are reduced due to the enhanced memory bandwidth and smaller column mux ratio. Although the delay reduction is negligible with $B_{IO} = 64$, DIMA-CNN still achieves a 2.4× energy improvement. Overall results indicate that DIMA-CNN achieves a 2.5× to 11.9× EDP reduction with $R = 50$ as compared to the conventional architecture with less

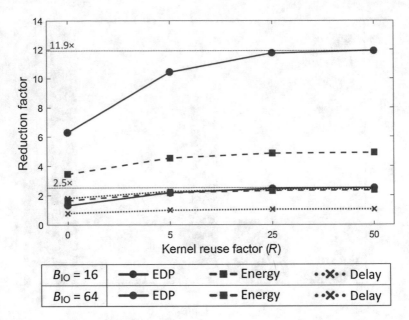

Fig. 5.12 Energy, delay, and EDP reductions over the conventional architecture (FX)

Table 5.3 Comparison with other DNN implementations on the MNIST dataset

Related works	Process (nm)	Network type	Throughput (K decisions/s)	Energy (nJ/decision)	EDP (pJ·s/decision)	Decision accuracy
[7]	28	5-layer perceptron	15	360	24	98%
[81]	65	LeNet-5	13	2463	184	>98%
[82]	65	Spiking NN	1650	162	0.1	90%
This work	65	LeNet-5	70	450	6	>97%[a]

[a]Median: 99%

than 1.3% degradation in recognition accuracy. That EDP improvement is less than the reported benefit 24.5× in [41] as practical hardware implementation constraints have been considered here.

As listed in a Table 5.3, DIMA-CNN achieves at least 4× smaller EDP compared to prior DNN implementations with comparable accuracy, e.g., >97%, for the MNIST dataset.

5.5   Sparse Distributed Memory (SDM)

There is much interest in exploring brain-inspired models of computation that can provide robust system behavior for inference applications while achieving high energy efficiency [83–86]. The sparse distributed memory (SDM) [87] (see

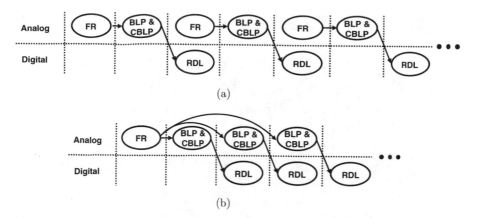

Fig. 5.6 Data-flow of DIMA-CNN: (**a**) without data ($\mathbf{w}_{nm}$) reuse, and (**b**) with data reuse

5.2.5   Sliding Window FM Register

Figure 5.7 describes data movement in the FM register and the input FIFOs during convolution. Accessing of $\mathbf{w}_{nm}$s in the bitcell array via the standard SRAM interface (see Fig. 2.3a) is delay and energy inefficient. Therefore, a matching portion of $\mathbf{x}$ is made available at every stride of the input window. Each FIFO stores K words, and K such FIFOs align a total of K^2 kernels. When the processing window slides horizontally (shown in red), the FM register provides new words $(X_{(1,6)}, \ldots, X_{(5,6)})$ to the input of each FIFO by shifting all the FIFO contents to the left. On the other hand, when the window slides vertically (shown in blue), new words $(X_{(6,1)}, \ldots, X_{(6,5)})$ are input to the Kth (right most) FIFO by shifting it K times to the left while all the other FIFOs' contents are transferred to the left neighboring FIFOs, i.e., the K words in the ith FIFO are transferred to $(i-1)$th FIFO.

5.3   Energy, Delay, and Functional Models of DIMA-CNN

This section provides behavioral, energy, and delay models for DIMA-CNN to be employed in system simulations in Sect. 5.4. In addition, the parameters for the models are validated with the measurement of the prototype IC from [32, 44] for the analog blocks, and with post-layout simulations for digital blocks, respectively.

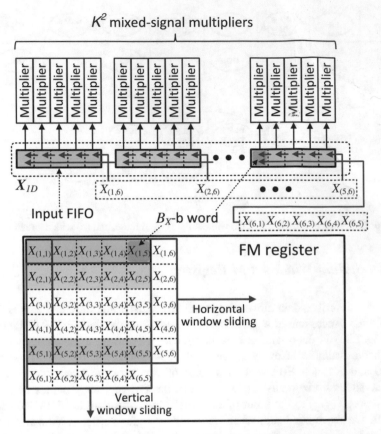

Fig. 5.7 FM register to enable window sliding and input FIFOs with $K = 5$, e.g., the green-marked right most FIFO contents $\{X_{(5,1)}, X_{(5,2)}, \ldots, X_{(5,5)}\}$ are replaced by $\{X_{(5,2)}, X_{(5,3)}, \ldots X_{(5,6)}\}$ after horizontal window sliding or by $\{X_{(6,1)}, X_{(6,2)}, \ldots, X_{(6,5)}\}$ after vertical sliding

5.3.1 Functional Model

DIMA is analog-intensive and therefore subject to a number of circuit-level non-idealities. Dominant among them are the following:

- non-linearity of functional READ process due to V_{BL}-dependent discharge path resistance R_{BL},
- impact of process variations on ΔV_{BL} and ΔV_m due to local transistor threshold voltage V_t-mismatch caused primarily by random dopant fluctuations,
- input offset of the analog differential comparators,
- charge loss on the node V_s due to the leakage current, leading to the inaccuracy on the node V_{in},

- mixed-signal multiplier offset (c/b in (5.5)) due to the non-ideal transistor switching behavior.

The non-linearity of functional READ in *B1)* is modeled by a polynomial fit of measured results in [32, 44] as $\Delta \widetilde{V}_{BL}(W) = \sum_{k=0}^{5} c_k W^k$, where $\Delta \widetilde{V}_{BL}(W)$ is a distorted version of $\Delta V_{BL}(W)$, and c_ks are the fitting parameters. The impact of process variations in *B2)* is modeled as a Gaussian distributed random variable (RV) as shown below:

$$\Delta \widehat{V}_{BL}(W) \sim \mathcal{N}(\Delta \widetilde{V}_{BL}(W), \sigma_{BL}^2(W)) \tag{5.6}$$

where $\sigma_{BL}^2(W)$ is the variance of $\Delta \widehat{V}_{BL}(W)$. The same modeling method is applied for the analog multiplier with $\sigma_m^2(V_{in})$, the variance of ΔV_m in (5.5). Monte Carlo SPICE simulations show that $\sigma_{BL}(W)$ and $\sigma_m(V_{in})$ are less than 12.5 and 6.5% of their average values, respectively. Note that the noise due to the variation is well averaged out during the aggregation of 128 BL processing outputs in the cross BL processing stage. The measurement results of prototype IC in [32] confirms this as the standard deviation of cross BL processing output across 128 rows is less than 1.1% of the average value.

The comparator offset in *(B3)* is modeled as:

$$V_{mux}(W) \sim \max(V_{BL}(W), V_{BLB}(W) + \eta) \tag{5.7}$$

where η is a zero-mean Gaussian distributed RV with 10 mV standard deviation [29].

The impact of leakage in *(B4)* on the node V_{in} as a function of the reuse number r is modeled as follows:

$$\widehat{V}_{in} = V_{in}e^{-\gamma(V_{in})r} \simeq V_{in}(1 - \gamma(V_{in})r) \tag{5.8}$$

where $\gamma(V_{in})$ is the discharge rate of V_{in} after each reuse. Provided that $r \ll 1/\gamma(V_{in})$, the impact of leakage can be approximated as shown in (5.8). SPICE simulations show that $\gamma < 0.05\%$ at the FF corner and at 85 °C, where leakage current is maximized. This also includes the kick-back noise from the toggling of switch *mult_on* to the node V_s. The parameter r is uniformly distributed, i.e., $r \sim U(1, R)$, where $R \ll 1/\gamma(V_{in})$ is the maximum allowed reuse number.

Figure 5.8 shows the measured results for the mixed-signal multiplier fabricated in a 65 nm process that is used to model the non-ideality *(B5)*. The model parameters in (5.5) can be written as $\alpha = ab$, and $\beta = c/b$, and can be fitted to the curve in Fig. 5.8 with $\alpha = 0.16$, and $\beta = -0.5$. The maximum modeling error is less than 2% of ΔV_m's dynamic range. The offset β results in a deviation from the ideal product $\alpha X V_{in}$, but since it is deterministic, it can be compensated via offline retraining of $\mathbf{w}_{nm}$ and C_n in (5.1) can compensate for it.

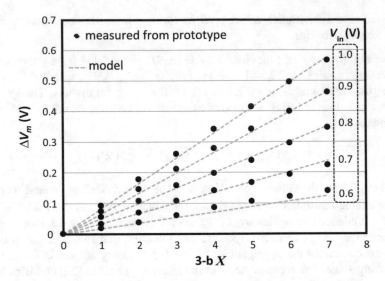

Fig. 5.8 Measured accuracy of the mixed-signal multiplier fabricated in the prototype IC of [32] compared with the model in (5.5) with $\alpha = ab = 0.16$ and $\beta = c/b = -0.5$

5.3.2 Delay and Energy Models

This section provides the delay and energy models for the conventional and proposed DIMA architectures. We focus on the C-layer, but these models can also be applied to the F-layers with slight modification. The conventional system generates N output FMs with a delay given by:

$$T_{\text{conv}} \approx \left\lceil \frac{MNK^2}{(B_{\text{IO}}/B_W)N_{\text{bank}}} \right\rceil T_{\text{read}} + \left\lceil MNK^2/N_{\text{mult}} \right\rceil N_{\text{mov}} T_{\text{mult}} \quad (5.9)$$

where $N_{\text{mov}} = (L - K + 1)^2$ is the number of window slidings across input FM, B_{IO} is the bit width of SRAM IO so that B_{IO}/B_W (e.g., 2–8) words are fetched per read access per bank, N_{mult} is the number of digital multipliers, and T_{read} and T_{mult} are the delays per the memory access and multiplication, respectively. The delay and energy for the aggregation via an adder tree is assumed to be negligible. The first and second terms in (5.9) correspond to the delays of SRAM read and subsequent convolution operations. Similarly, DIMA-CNN generates N output FMs with a delay given by:

$$T_{\text{DIMA}} \approx \left\lceil \frac{MNK^2}{(N_{\text{bank}}N_{\text{COL}}/2)} \right\rceil \times (\lceil N_{\text{mov}}/R \rceil T_{\text{FR}} + N_{\text{mov}} T_{\text{BLP}}) \quad (5.10)$$

where T_{FR} and T_{BLP} are the delays of functional READ and BL processing stages, respectively. DIMA can access and process $N_{\text{COL}}/2$ (e.g., 128) words per bank in parallel, where factor 2 in the denominator arises from the use of sub-ranged read (see Sect. 2.4.1). This parallelism creates a significant throughput enhancement, as we can see by comparing the first terms in (5.10) and (5.9).

The energy consumption of the conventional architecture in generating N output FMs is given by:

$$E_{\text{conv}} \approx MNK^2 E_{\text{read}} + MNN_{\text{mov}} E_{\text{reg}} + MNK^2 N_{\text{mov}} E_{\text{mult}} + P_{\text{leak}} T_{\text{conv}} \quad (5.11)$$

where E_{read}, E_{mult}, and P_{leak} represent the single word SRAM read energy, digital multiplier energy, and SRAM leakage power consumption, respectively. It is assumed that a deep-sleep mode is enabled to minimize the leakage during standby through the use of techniques such as power gating or lowering of the supply voltage for the bitcell array [46]. Here, E_{reg} is the register energy to: (1) fetch five words per window sliding from the FM register, (2) shift the contents of input FIFOs per window sliding in Fig. 5.7, and (3) write back a word of the output FM into the FM register after four such window shifting and sub-sampling. Similarly, the energy consumption of DIMA is given by:

$$E_{\text{DIMA}} \approx MNK^2 \lceil N_{\text{mov}}/R \rceil E_{\text{FR}} + MNN_{\text{mov}} E_{\text{reg}}$$
$$+MNK^2 N_{\text{mov}} E_{\text{BLP}} + P_{\text{leak}} T_{\text{DIMA}} \quad (5.12)$$

where E_{FR} and E_{BLP} are the energy consumed for the functional READ and BL processing operations of a single word, respectively. They are significantly smaller (by roughly $10\times$) than E_{read} and E_{mult} due to the DIMA's read and processing with low-voltage swing. These models are employed in Sect. 5.4 to demonstrate the benefits of the proposed architecture.

The parameter values $T_{\text{FR}} = 7$ ns and $T_{\text{BLP}} = 17$ ns in (5.9)–(5.12) are validated via the prototype IC in [32] through the measurement functional READ and BL processing accuracies within the given delay. The conventional SRAM read of the prototype IC in [32] takes 9 ns ($= T_{\text{read}}$) per single read access. However, it could be improved if a self-timed controller is employed. Thus, for the conventional system, $T_{\text{read}} = 4$ ns is assumed to be conservative. The digital blocks in the conventional system are also synthesized to have the same delay ($T_{\text{mult}} = 4$ ns) as T_{read} to enable two-stage pipeline between the dot product processing and the SRAM access in F-layers. The values of E_{FR}, E_{read}, E_{BLP}, and P_{leak} are also validated by the comparison of the measured total energy of prototype IC in [32] and that from the post-layout simulations. Other parameters for digital blocks such as E_{mult} and E_{reg} are estimated from post-layout SPICE simulations of synthesized blocks. The delay, energy, and behavioral model parameters are summarized in Table 5.1.

The F-layers are implemented much like the C-layers except that neither the sliding window nor the reuse of functional READ outputs is required. Therefore, the parameters of the F-layer model need to be set accordingly, e.g., $r = 0$ in (5.8), and $N_{\text{mov}} = 1$, $R = 1$ in (5.9)–(5.12).

Table 5.1 Delay, energy, and behavioral model parameters

Parameter	Values	Parameter	Values
T_{FR}, T_{read}	7 ns, 4 ns	T_{BLP}	17 ns
T_{pulse}	2 ns	T_{mult}	4 ns
E_{BLP}, E_{mult}	0.08 pJ, 0.9 pJ	E_{FR}, E_{read}	0.5 pJ, 5.2 pJ
E_{reg}, P_{leak}	4 pJ, 2.4 nW	C_S,C_c,C_m	25, 25, 100 fF
$\sigma(W)$	7–12.5% of avg.	$\sigma_m(V_{in})$	2.5–6.5% of avg.
α, β	0.16, −0.5	γ	<0.05%
$c_0,\ldots,c_5$	$-0.04, 0.97, -0.14, 4.7\times10^{-2}$ $-5.3\times10^{-3}, 2.5\times10^{-4}, -4.3\times10^{-6}$		

5.4 Simulation and Results

This section describes the system configuration and simulated results for the proposed DIMA-CNN architecture in terms of its energy, delay, and inference accuracy. As shown in Sect. 5.3.1, the accuracy analysis needs to incorporate the stochastic nature of DIMA, and thus requires a sufficiently large number of simulations. Therefore, we employ LeNet-5 [75] for hand written digit recognition with the MNIST database [64].

5.4.1 Architecture Configurations

LeNet-5 [75] has six layers: C-layers C1 and C3, F-layers F5 and F6, and S-layers S2 and S4. The design parameters of DIMA-CNN are summarized in Table 5.2. Bit precision of $B_W = 8$ and $B_X = 6$ are chosen to obtain negligible accuracy degradation compared to a floating point realization. The DIMA-CNN with 64 KB SRAM ($N_{bank}N_{ROW}N_{COL} = 4 \times 512 \times 256$) and a total of 2 KB registers are employed, where 20 $\mathbf{w}_{mn}$s ($N_{COL}N_{bank}/(2K^2) \approx 20$ with $K = 5$) are stored per word-row and processed per read access cycle. The FM registers for C-layers are designed to support the vertical and horizontal window movements with kernel size $K = 5$. The SRAM, BL processing/cross BL processing blocks, and registers take 79.8%, 8.6%, and 11.6% of overall layout area, respectively.

We assume the conventional architecture comprises a conventional SRAM instead of a functional READ, and digital dot-product computations instead of BL processing and cross BL processing blocks. For the conventional architecture, the IO bit-widths (B_{IO}) per SRAM bank of 16, 32, and 64 bits are considered corresponding to a column mux ratio $L = 4$, 8, and 16, respectively. A $K^2 (= 25)$ dimensional digital dot product block is synthesized and post-layout simulated in the same (65 nm) process. For fair comparison, we assume that the conventional architecture has seven digital dot product blocks as the area occupied by these blocks is approximately equal to the BLP and CBLP blocks in DIMA-CNN. The DIMA-CNN achieves

Table 5.2 Design parameters of LeNet-5

Parameter	Values	Parameter	Values
V_{DD}	1 V	V_{PRE}	1 V
Input L	32	K	5
B_W, B_X	8, 6	N_{mult}	175
R	0–200	B_{IO}	16–64
$(N)_{bank,row,col}$	4, 512, 256		
N	C1: 6 , C3: 16, F5: 120, F6: 10		

higher parallelism, e.g., the number of BL processing blocks $N_{COL}N_{bank}/2 = 512$, than the conventional architecture with $N_{mult} = 175$ (= 25×7) within the same area due to the pitch-matched BL processing layout as shown in Fig. 5.5c. The energy and delay of the BL processing blocks are estimated via post-layout simulations.

5.4.2 Recognition Accuracy

The error rates (ϵ_{det}) are measured on MNIST test dataset [64] via system simulations for the following architectures:

- FL: floating-point simulations.
- FX: the conventional (SRAM+digital processor) architecture with error-free fixed-point computations (B_W and B_X).
- DIMA: DIMA-CNN with fixed-point computations that use behavioral models (5.5)–(5.8) without resorting to retraining.
- DIMA-TR: DIMA-CNN with fixed-point coefficients retrained accounting for the non-linearity of functional READ process and the offset of charge-recycling mixed-signal multiplier.

Retraining (<10 iterations) employs the behavioral model of the charge-recycling multiplier in (5.5), and is applied after the standard training process (100 iterations). The ϵ_{det}s of DIMA and DIMA-TR are obtained via Monte Carlo simulations with 400 iterations at each value of R. DIMA and DIMA-TR have the same statistical behavior as DIMA-TR is retrained to compensate for only deterministic non-idealities. Figure 5.9 plots the ϵ_{det}s of above architectures for different values of the reuse factors R. The floating-point simulations (FL) and the conventional architecture (FX) achieve a constant $\epsilon_{det} = 0.8$ and 0.97% across R, respectively. In contrast, the ϵ_{det} of DIMA increases with R because of the leakage current as described in (5.8) resulting in a median ϵ_{det} of 1.7% with $R = 800$.

However, the median ϵ_{det} of DIMA-TR reduces by up to 0.2% due to retraining. The ϵ_{det} of DIMA-TR increases negligibly, with a median $\epsilon_{det} < 1.3\%$ (worst-case, $\epsilon_{det} < 2.3\%$) for $R < 200$.

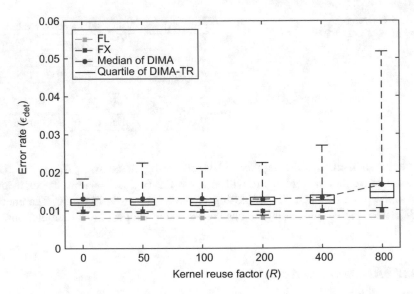

Fig. 5.9 Error rate (ϵ_{det}) vs. reuse factor (R)

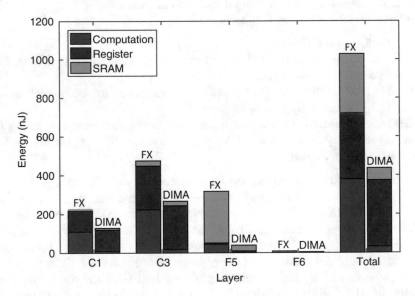

Fig. 5.10 Energy breakdown per layer with $B_{IO} = 16$ and $R = 50$

5.4.3   Energy and Delay

The energy and delay of the conventional architecture and DIMA-CNN are estimated based on (5.9)–(5.12). Figure 5.10 shows the energy breakdowns across the
C- and F-layers but not the S-layers, which has a negligible contribution to the total

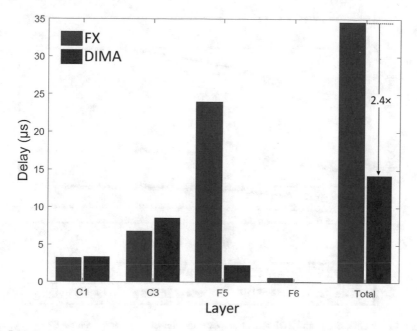

Fig. 5.11 Delay of each layer with $B_{IO} = 16$ and $R = 50$ for the conventional architecture (FX) and DIMA

energy and delay. The DIMA-CNN architecture saves energy in every layer because of the low-swing computation in functional READ, BL processing, and cross BL processing stages. Specifically, DIMA significantly reduces the energy in F5 layer, where the highest data access volume is required compared to the other layers. The overall system energy is reduced by $4.9\times$ when compared to the conventional architecture, achieving an energy efficiency of 436 nJ/decision.

As shown in Fig. 5.11, DIMA-CNN has similar or slightly higher delay in C-layers as compared to the conventional architecture since the proposed mixed-signal multiplier is slower than digital multiplication ($T_{BLP} > T_{mult}$). However, DIMA-CNN achieves a significant throughput improvement in the memory-intensive F5 layer because of the highly parallel functional READ process. Overall, DIMA-CNN achieves approximately a $2.4\times$ delay reduction at a decision latency of 14.3 μs/decision.

Figure 5.12 shows that both energy and delay benefits over the conventional architecture increase with the resue factor R, achieving up to $11.9\times$ EDP reduction. The improvement in EDP saturates at $R = 50$ as the contribution of functional READ to the total energy and delay becomes negligible when $R > 50$. As B_{IO} increases, the energy and delay of the conventional architecture are reduced due to the enhanced memory bandwidth and smaller column mux ratio. Although the delay reduction is negligible with $B_{IO} = 64$, DIMA-CNN still achieves a $2.4\times$ energy improvement. Overall results indicate that DIMA-CNN achieves a $2.5\times$ to$11.9\times$ EDP reduction with $R = 50$ as compared to the conventional architecture with less

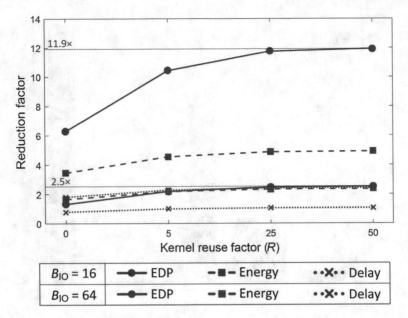

Fig. 5.12 Energy, delay, and EDP reductions over the conventional architecture (FX)

Table 5.3 Comparison with other DNN implementations on the MNIST dataset

Related works	Process (nm)	Network type	Throughput (K decisions/s)	Energy (nJ/decision)	EDP (pJ·s/decision)	Decision accuracy
[7]	28	5-layer perceptron	15	360	24	98%
[81]	65	LeNet-5	13	2463	184	>98%
[82]	65	Spiking NN	1650	162	0.1	90%
This work	65	LeNet-5	70	450	6	>97%[a]

[a]Median: 99%

than 1.3% degradation in recognition accuracy. That EDP improvement is less than the reported benefit 24.5× in [41] as practical hardware implementation constraints have been considered here.

As listed in a Table 5.3, DIMA-CNN achieves at least 4× smaller EDP compared to prior DNN implementations with comparable accuracy, e.g., >97%, for the MNIST dataset.

5.5   Sparse Distributed Memory (SDM)

There is much interest in exploring brain-inspired models of computation that can provide robust system behavior for inference applications while achieving high energy efficiency [83–86]. The sparse distributed memory (SDM) [87] (see

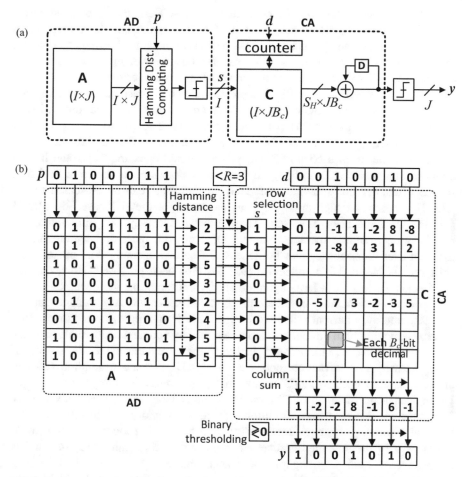

Fig. 5.13 Sparse distributed memory (SDM) with *address decoder* (AD) and *counter array* (CA): (a) architecture (S_H: number of selected rows), and (b) example of SDM operation ($I = 8$, $J = K = 7$, and $S_H = 3$)

Fig. 5.13) is one such computational model of the human brain. An SDM can be trained to remember sparse data vectors and retrieve them when presented with noisy or incomplete versions of the stored vectors. That is similar to the human brain's ability to associate a related memory given noisy sensory input by conceptualizing/categorizing incomplete information [88].

Being a memory array, the SDM input is a 2-tuple (p, d), where p and d are the J-bit address and K-bit data, respectively (we assume $J = K$ in the rest of this section). In an SDM, data vectors d are first stored (WRITE operation). The *address decoder* (AD) projects the J-bit address vector p onto a higher I-dimensional ($I \gg J$) space, and then uses this high-dimensional representation s of p as the decoded address in the *counter array* (CA), where d is stored in a distributed fashion. In the

READ mode, the address p is first decoded by the AD, and the decoded address s is used to retrieve the stored data from the CA. The sparse and distributed nature of data processed and stored in an SDM provides inherent robustness to noise or imprecision in the input data. The SDM can also be employed in an auto- or hetero-associative mode to achieve even greater robustness to data errors.

However, a straightforward SDM implementation will consume much energy and will be slow because the SDM operates in a high-dimensional (hyper-dimensional) space [87], e.g., typical SDM parameters are $I = 2 \times 10^3$ to 10^6, $J \geq 256$, $B_c \geq 5$, where B_c is a bit precision of each counter in the CA [88, 89]. Such an implementation in a 65 nm CMOS process would consume 77 µJ and have a delay of 2 ms per READ. In fact, the dominant (about 80% as shown in Sect. 5.8) source of energy consumption and delay in the SDM can be attributed to the AD. Hence, several high-throughput architectures for the AD based on SRAM and DRAM have been proposed. These achieve speed-up by parallelizing the AD by using multiple memory blocks [90]. However, these architectures suffer from an inter-block throughput bottleneck. To remove memory read operation, a shift register-based AD architecture [91] has also been proposed. However, this architecture suffers from large dynamic energy consumption and occupies a large area compared to memory-based architectures. Mixed-signal AD implementations [91, 92] employ a current mirror to evaluate the Hamming distances in parallel, thereby achieving high throughput. However, the large content addressable memory bitcell dimension (i.e., 11 transistors, including the current mirror) results in a loss of storage density, and the bias currents result in high DC power consumption. One can also implement a design of an SDM by employing resistive memory devices [89].

Implemention of an the SDM model requires large storage capacity closely integrated with computation. Traditional processor-memory architectures separate low-swing memory storage functionality from high-swing logic. That separation exists even in the so-called "processor-in-memory architecture" [12, 14], and is the source of both a throughput bottleneck and energy consumption. In fact, conventional architectures fail to exploit an important feature of the SDM [88]: its ability to compensate for hardware noise/errors in addition to noise/errors in the input data. In contrast, DIMA can exploit SDM's intrinsic error-tolerance to compensate for its analog non-idealities while alleviating the energy and latency costs of a traditional processor-memory interface.

5.5.1 SDM Functionality

Figure 5.13 shows that the SDM accepts as input a 2-tuple (p, d) with a J-bit address p and J-bit data d. The SDM architecture (see Fig. 5.13a) includes: (1) a J-bit *address decoder* AD to evaluate the Hamming distance between p and the I, J-bit addresses are stored in a $I \times J$ memory array $\mathbf{A}$ in the AD; and (2) a *counter array* CA with a counter and a memory array $\mathbf{C}$ to store the IJ B_c-bit counts.

WRITE Operation During the WRITE operation, the AD generates an I-bit decoded row address $s = [s_1, s_2, \ldots, s_i]$ (see Fig. 5.13), as follows:

$$s_i = \text{sgn}\{R - \sum_{j=1}^{J}(a_{ij} \oplus p_j)\}, \quad (i = 1, 2, \ldots, I) \tag{5.13}$$

$$\text{sgn}(x) = \begin{cases} 1, & \text{if } x \geq 0 \\ 0, & \text{otherwise,} \end{cases}$$

where $\oplus$ is the binary EXOR operator, $a_i = [a_{i1}, a_{i2}, \ldots, a_{ij}]$ is the ith address stored in A, $p = [p_1, p_2, \ldots, p_j]$ is the input address, and R is a user-defined radius/threshold.

The decoded row address s and the data d are employed in the CA to update the count, as follows:

$$c_{ij} \leftarrow \begin{cases} c_{ij} + s_i, & \text{if } d_j = 1 \\ c_{ij} - s_i, & \text{otherwise,} \end{cases} \tag{5.14}$$

where $d = [d_1, d_2, \ldots, d_j]$. Note that the contents of C are updated only when $s_i = 1$, i.e., only the selected rows are updated.

READ Operation During the READ operation, AD generates the row address s in the same manner as in the WRITE operation described by (5.13). Then, the SDM output is read out as:

$$y_j = \text{sgn}(s \cdot c_j), \quad (j = 1, 2, \ldots, J) \tag{5.15}$$

where c_j is the jth column vector of C, and $y = [y_1, y_2, \ldots, y_j]$ is the output word.

5.5.2 Associative Memory

In associative memories, the data read are the stored data that are most strongly associated with the contents of the input rather than a specific address. There are two types of associative memories: (1) auto-associative memory, and (2) hetero-associative memory. The SDM can operate in both modes of associative recall, and when it does, the SDM exhibits even stronger robustness to noise/errors in data.

In the auto-associative mode, one can train the SDM is trained by selecting its input 2-tuple (p, d) from the training set $S_t = \{(t_1, t_1), (t_2, t_2), \ldots\}$, i.e., both the address p and the data d are assigned the same value [93]. On the other hand, in the hetero-associative mode, one trains the SDM by selecting its input 2-tuple (p, d)

from the training set $S_t = \{(t_{11}, t_{12}), (t_{21}, t_{22}), \ldots\}$, i.e., p and d are assigned different values.

During the classification/decision-making phase, in both associative modes, the SDM is operated in an iterative manner where initially $p = l$, where l is a noisy/incomplete version of the stored data. Then, in subsequent iterations, p is set to the current output of the SDM. Thus, the classification phase of the SDM is described as follows:

$$p[n] = \begin{cases} l, & \text{if } n = 1 \\ \mathbf{y}[n-1], & \text{if } n > 1 \end{cases} \tag{5.16}$$

where n is the time index, and l is an initial input. The output $\mathbf{y}[n]$ converges to an error-free/closest version of l that was stored during the training phase. The SDM's auto- and hetero associative modes can be interpreted as the human brain's ability to extract a pattern from noise and locate the next pattern in a certain sequence given the current pattern [88].

5.5.3 The Conventional SDM Architecture

The conventional SDM architecture is multi-block [90] (see Fig. 5.14a) in order to enhance throughput. The multi-block SDM architecture comprises M blocks ($\mathbf{B}_1$, $\mathbf{B}_2$,...,$\mathbf{B}_m$) that operate in parallel, where each block has its own *address decoder* AD_m with memory array $\mathbf{A}_m$ of size $(I/M) \times J$ bits, a *counter array* CA_m with memory array $\mathbf{C}_m$ of size $(I/M) \times (JB_c)$ bits, and decoded row address s_m ($m = 1, \ldots, M$). The memory array $\mathbf{A}$ in AD is implemented via SRAMs for high throughput, while the memory array $\mathbf{C}$ in CA is implemented using DRAM, Flash, and PRAM, in order to achieve high storage densities.

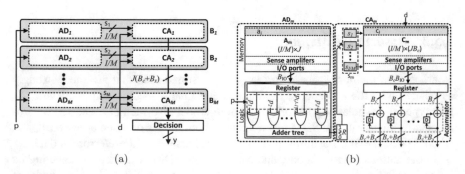

(a)   (b)

Fig. 5.14 The conventional SDM implementation: (**a**) an M-parallel block architecture, and (**b**) architecture of a single block

The architecture of each block (see Fig. 5.14b) indicates that AD_m computes the Hamming distance between the input address p and (I/M) stored addresses in $\mathbf{A}_m$, while CA_m generates a partial sum, which is then accumulated and thresholded during the READ operation. The each partial sum requires an additional B_X bits in addition to B_c bits per single counter in order to prevent overflow, where B_X depends upon the sparsity of the stored data and R. Thus, the multi-block SDM architecture requires $J(B_c + B_X)$ global BLs (GBLs) per block to transfer partial sums to the decision block.

The conventional architecture in Fig. 5.14a has a number of drawbacks. Key among them are the following:

1. The Hamming distance computation in the AD_m requires access to all the memory locations. The throughput of AD_m is limited by the SRAM read out bandwidth. In particular, multiple read out cycles are required in order to read a single a_i in a conventional memory and processor architecture due to column-muxing requirements.
2. The additional digital blocks in the AD, such as the adder tree and EXOR gates, lead to energy consumption and area overhead.
3. Routing of the GBLs in the CA_m is made difficult because of their large number ($J(B_c + B_X)$ per block), and because of the small bitcell area ($4F^2$, F: $1/2$ of BL pitch) due to the use of high-density memories [94, 95].

Section 5.6 describes how those drawbacks of the conventional architecture can be overcome via DIMA.

5.6 DIMA-Based SDM Architecture (DIMA-SDM)

In this section, a DIMA-based SDM (DIMA-SDM) is proposed (see Fig. 5.15a) to address the drawbacks of conventional architectures listed in Sect. 5.5.3. In particular, DIMA-SDM employs the following key techniques:

- The AD was designed using DIMA (DIMA-AD) (see Fig. 5.15b) in order to overcome the bandwidth limitation and eliminate the use of digital logic.
- The CA was implemented using a hierarchical binary decision (HBD) technique (CA-HBD) as shown in Fig. 5.15c, in order to minimize the routing overhead of GBLs.

5.6.1 DIMA-Based Address Decoder (DIMA-AD)

The proposed DIMA-AD generates the Hamming distance per (5.13) via a three-step process: (1) the FR process generates BL voltages V_{BL} and V_{BLB} that are proportional to the sum $a_{ij} + p_j$ over the field of real numbers, followed by; (2)

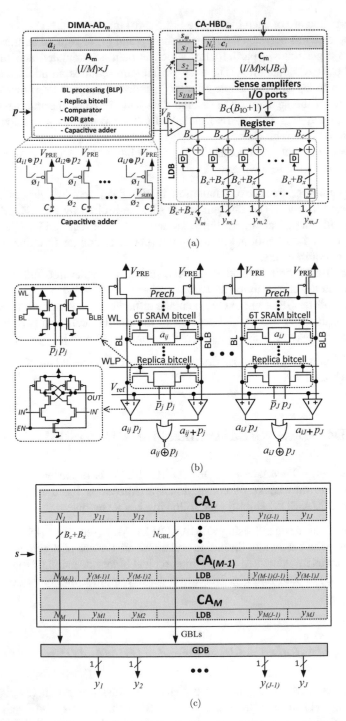

Fig. 5.15 Proposed SDM architecture (DIMA-SDM): (**a**) architecture of single block, (**b**) AD with DIMA (DIMA-AD) including deeply embedded mixed signal processing units, and (**c**) CA with hierarchical binary decision (CA-HBD) (N_{GBL}: number of GBLs)

Fig. 5.16 Multi-row read functional read (FR) for EXOR operation in DIMA-AD when $a_{ij} = 0$

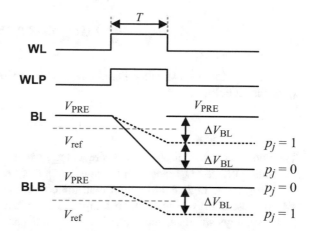

the use of BLP to compute $a_{ij} \oplus p_j$; and (3) finally the Hamming distance (via a capacitive adder). These steps are described next.

The FR step begins with the simultaneous application of access pulses to the rows storing a_{ij} and p_j such that the pulse width $T \ll R_{BL}C_{BL}$, where $R_{BL}C_{BL}$ is the RC time constant of BL/BLB [29]. That results in a BL/BLB voltage (see Fig. 5.16) given by:

$$V_{BL} = V_{PRE} - (\overline{a}_{ij} + \overline{p}_j)\Delta V_{BL} \qquad (5.17)$$

$$V_{BLB} = V_{PRE} - (a_{ij} + p_j)\Delta V_{BL} \qquad (5.18)$$

where $\Delta V_{BL} = V_{PRE}(T/R_{BL}C_{BL})$. A replica bitcell is employed to avoid writing **p** into the main array **A** (see Fig. 5.15b).

The second step (BLP) begins with the BL/BLB provided as inputs to differential comparators [66] sized to fit within a single bitcell pitch with an appropriately selected reference voltage $V_{ref} = V_{PRE} - \Delta V_{BL}/2$. Doing so results in binary-valued comparator outputs:

$$X_{BL} = a_{ij} p_j = \text{sgn}(V_{\text{diff,BL}}) = \text{sgn}\{0.5 - (\overline{a}_{ij} + \overline{p}_j)\} \qquad (5.19)$$

$$X_{BLB} = \overline{a_{ij} + p_j} = \text{sgn}(V_{\text{diff,BLB}}) = \text{sgn}\{0.5 - (a_{ij} + p_j)\}$$

A NOR2 gate that combines the comparator outputs generates $a_{ij} \oplus p_j$ as follows:

$$a_{ij} \oplus p_j = \overline{\text{sgn}\{0.5 - (a_{ij} + p_j)\} + \text{sgn}\{0.5 - (\overline{a}_{ij} + \overline{p}_j)\}} \qquad (5.20)$$

Next, a J-bit capacitive adder (see Fig. 5.15a) accepts $a_{ij} \oplus p_j$ from the NOR2 gate output and employs charge redistribution to compute the summation in (5.13) as follows:

$$V_{\text{sum},i} = \frac{1}{J} \sum_{j=1}^{J} (1 - a_{ij} \oplus p_j) V_{\text{PRE}} \tag{5.21}$$

The last step involves an analog comparator that generates the decoded address bit s_i as shown below:

$$s_i = \text{sgn}(V_{\text{sum},i} - V_R) \tag{5.22}$$

The above sequence of operations is repeated I/M times.

Thus, DIMA-AD reads a_i in a single read cycle (single precharge) and has $\approx J/B_{\text{IO}}$ times higher throughput than the conventional AD. In addition, DIMA-AD is more energy-efficient than a digital implementation because the capacitive adder employs small capacitances (i.e., $C = 10\,\text{fF}$) and requires a simple switching operation.

5.6.2 Counter Array Using Hierarchical Binary Decision (CA-HBD)

The proposed CA-HBD architecture minimizes the inter-block data transfer as shown in Fig. 5.15c, where GDB and LDB are global and local decision blocks, respectively. The CA-HBD architecture requires recording of the row access count N_i, i.e., the number of accesses to each physical address a_i in **A** during the WRITE operation. The row access count N_i is recorded in an additional column in the CA.

During the READ operation, the LDB of the mth block generates a local binary decision $y_{\text{m,j}}$ and N_m as follows:

$$N_m = \sum_{i \in H_m} N_i \tag{5.23}$$

$$y_{m,j} = \text{sgn}\left(\sum_{i \in H_m} c_{ij} \right) \tag{5.24}$$

where H_m is the set of row indices in the mth block CA_m that were selected during the READ operation, and N_m represents the sum of the row access counts for these rows. Finally, the GDB (see Fig. 5.17) generates the final SDM output bit y_j as follows:

$$y_j = \text{sgn}\left\{ \sum_{m=1}^{M} \text{sign}(y_{m,j}) N_m \right\}, \text{ where} \tag{5.25}$$

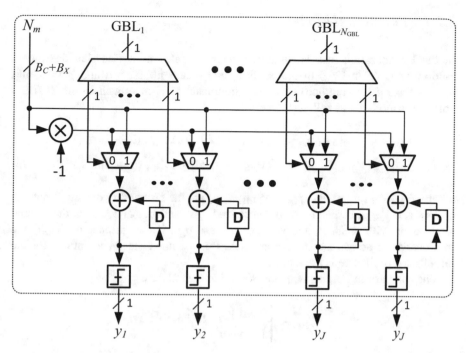

Fig. 5.17 Global decision block (GDB) to incorporate local decisions ($y_{m,j}$) with impact factor N_m

$$\text{sign}(x) = \begin{cases} 1, & \text{if } x > 0 \\ -1, & \text{otherwise} \end{cases}$$

Thus, the GDB weights CA_m's contribution $y_{m,j}$ by N_ms in order to assign more weight to those blocks that were accessed more frequently during the WRITE phase. In that manner, the LDB transmits compressed information to the GDB, as $y_{m,j}$s are binary numbers. Thus, J-bits are required instead of $J(B_c + B_X)$-bits, as shown in Fig. 5.15c, thereby minimizing the delay and the energy penalty for the data transfer.

5.7   Energy, Delay, and Functional Models of DIMA-SDM

Functional models that capture the non-ideal behavior of DIMA-AD are essential for predicting system performance. The analog-intensive FR and BLP operations of the DIMA-AD are intrinsically vulnerable to various sources of noise due to its low-SNR operation. The dominant sources of noise in the DIMA-AD are: (1) the local transistor threshold voltage V_t-variation across bitcells caused by random dopant fluctuations; and (2) input offset of the analog comparator.

5.7.1 Functional Model

In the FR operation, the V_t variations were modeled as a Gaussian distributed random variable in [29]. In this section, two binary numbers a and p (we omit indices i and j for simplicity) are FR. The impact of V_t-mismatch on the BL/BLB voltages is modeled as follows:

$$f_{V_{BL}}(V_{BL}; a, p) = \mathcal{N}(V_{PRE} - (\overline{a} + \overline{p})\Delta V_{BL}, (\overline{a} + \overline{a})\sigma_{cell}^2)$$

$$f_{V_{BLB}}(V_{BLB}; a, p) = \mathcal{N}(V_{PRE} - (a + p)\Delta V_{BL}, (a + p)\sigma_{cell}^2) \qquad (5.26)$$

where $f_{V_{BL}}(V_{BL}; a, p)$ and $f_{V_{BLB}}(V_{BLB}; a, p)$ are the probability density functions of V_{BL} and V_{BLB}, respectively, parametrized by a and p. $\mathcal{N}(\mu, \sigma^2)$ is the normal distribution with mean μ, and variance σ^2, and σ_{cell}^2 is the variance of ΔV_{BL} due to V_t variation across the storage array **A**. It is assumed that V_t variations for the bitcells and replica cells are identical.

The comparator outputs X_{BL} and X_{BLB} are obtained as:

$$X_{BL} = \begin{cases} 0 & \text{if } V_{BL} < V_{ref} + V_{OS} \\ 1 & \text{otherwise} \end{cases}$$

$$X_{BLB} = \begin{cases} 0 & \text{if } V_{BLB} < V_{ref} + V_{OS} \\ 1 & \text{otherwise} \end{cases}$$

$$f(V_{OS}) = \mathcal{N}(0, \sigma_{comp}^2) \qquad (5.27)$$

where an input offset voltage (V_{OS}) of the comparator is modeled as a zero-mean Gaussian random variable with variance σ_{comp}^2.

The charge injection noise in the switches and thermal noise/mismatch of capacitors in the capacitive adder are made negligible by ensuring $C > 10\,\text{fF}$ [96]. The single comparator at the output of capacitive adder can be designed to have a small input offset by using large transistor sizes and calibration techniques.

The behavioral models in this section are validated in Sect. 5.8.

5.7.2 Delay and Energy Models

The delay per READ of the conventional SDM and the DIMA-SDM are described as follows:

$$T_{SDM} = T_{AD} + T_{CA} \qquad (5.28)$$

$$T_{AD} = (I/M)(J/B_{IO})T_{read}$$

$$T_{CA} = S_{H,max}(J/B_{IO})T_{read} + M\lceil J(B_c + B_X)/N_{GBL}\rceil T_{GBL}$$

$$T_{\text{DIMA-SDM}} = T_{\text{DIMA-AD}} + T_{\text{CA-HBM}} \tag{5.29}$$

$$T_{\text{DIMA-AD}} = (I/M)T_{\text{read}}$$

$$T_{\text{CA-HBM}} = S_{\text{H,max}}(J/B_{\text{IO}})T_{\text{read}} + M\lceil J/N_{\text{GBL}}\rceil T_{\text{GBL}}$$

where T_{AD} ($T_{\text{DIMA-AD}}$) and T_{CA} ($T_{\text{CA-HBM}}$) are the delay for AD (DIMA-AD) and CA (CA-HBD). In addition, T_{read} is the delay for a single read access for memory arrays **A** and **C**, T_{GBL} is the delay in transferring a single bit via the GBLs, N_{GBL} is the number of GBLs, and $S_{\text{H,max}}$ is the maximum number of selected addresses per block. It is assumed that all other blocks are operating in parallel while the memories are being accessed. Hence, the delay of blocks such as the logic blocks in the conventional AD and the capacitive adder in DIMA-AD are not included in (5.28) and (5.29). The factors (I/M) and (J/B_{IO}) in T_{AD} are equal to the number of rows and the number of read outs per row, respectively, in AD_m. The partial sums per block are transferred serially through N_{GBL} GBLs, thus requiring $\lceil J(B_{\text{c}} + B_X)/N_{\text{GBL}}\rceil$ cycles.

The throughput enhancement of DIMA-SDM over SDM derives from: (1) $J/B_{\text{IO}} \geq 4$ in T_{AD}; and (2) $B_{\text{c}} + B_X \geq 8$ in T_{CA}. The delay models in (5.28) and (5.29) are plotted in Fig. 5.18, where it is assumed that T_{read} and T_{GBL} take two clock cycles. DIMA-SDM demonstrates a $25\times$ smaller delay than SDM with $B_{\text{IO}} = 8$ because of the high bandwidth of DIMA-AD when $M = 4$. The benefit of HBD at $M = 2048$ is evident as there is a $3.2\times$ additional delay reduction as compared to DIMA-SDM without HBD. The energy consumption per READ of the conventional SDM and the DIMA-SDM are modeled as follows:

$$E_{\text{SDM}} = E_{\text{AD}} + E_{\text{CA}} \tag{5.30}$$

$$E_{\text{AD}} = I[(J/B_{\text{IO}})(E_{\text{PRE}} + E_{\text{leak}}) + JE_{\text{SA}} + E_{\text{logic}}]$$

Fig. 5.18 Normalized delay of single READ operation with $B_{\text{IO}} = 8$–64 ($B_{\text{IO}} : J = 1 : 4$–32)

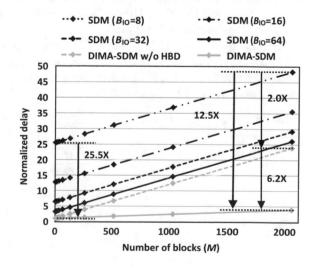

$$E_{CA} = S_H B_c[(J/B_{IO})(E_{PRE} + E_{leak}) + J E_{SA}]$$

$$E_{PRE} = J C_{BL} \Delta V_{BL} V_{PRE}$$

$$E_{leak} = I J P_{leak\_cell} T_{read}$$

$$E_{DIMA\text{-}SDM} = E_{DIMA\text{-}AD} + E_{CA\text{-}HBM} \qquad (5.31)$$

$$E_{DIMA\text{-}AD} = I(2E_{PRE} + E_{leak} + 2J E_{comp} + E_{a\_add})$$

$$E_{CA\text{-}HBM} < E_{CA}$$

where E_{AD} ($E_{DIMA\text{-}AD}$) and E_{CA} ($E_{CA\text{-}HBM}$) are the energy consumptions of the AD (DIMA-AD) and CA (CA-HBD), respectively. E_{PRE} and E_{leak} are the energy consumptions of the precharge and bitcell leakage for the entire memory array $\mathbf{A}$, respectively, and E_{SA} (E_{comp}) is for a single unit of the sense amplifier (analog comparator). P_{leak_cell} is the leakage power of each bitcell. The energy consumptions of the analog capacitive adder in the DIMA-AD and the logic blocks in the conventional AD per single Hamming distance computation are denoted by $E_{a\text{-}add}$ and E_{logic}, respectively. Energy consumptions from other blocks, such as WL drivers [97] and CA's decision blocks, are assumed to be negligible. It is assumed that the AD and CA can be placed into a deep sleep mode independently [46]. Note that $E_{AD} \gg E_{CA}$ because $I \gg S_H B_c$. The $E_{DIMA\text{-}AD}$ has a scaling factor of two for the first and third terms as DIMA-AD reads a_{ij} and p_j, and employs two comparators per bitcell column.

The energy efficiency of DIMA-AD derives from the fact that: (1) the first term in E_{AD} and $E_{DIMA\text{-}AD}$ is the largest, and $J/B_{IO} \geq 4$, (2) $E_{a\_add} \ll E_{logic}$, as the capacitances in the capacitive adder are very small, e.g., 10 fF, and the capacitive adder requires only simple switching operations; and (3) the leakage energy in $E_{DIMA\text{-}AD}$ is smaller than that in E_{AD} because the high-throughput (see delay models (5.28) and (5.29)) of DIMA-SDM permits it to be placed into a deep sleep mode much quicker than SDM [46].

The energy models (5.30) and (5.31) using typical design parameters from Table 5.4 are plotted in Fig. 5.19. The component values of (5.30) and (5.31) were obtained from Sect. 5.7.2. Figure 5.19 indicates that DIMA-SDM achieves energy reductions of $2.1\times$ to $12.4\times$ over SDM.

Table 5.4 Design parameters for SDM

Parameter	Value	Parameter	Value
$V_{DD}(= V_{PRE})$	1 V	M	4–2048
I/M	512	J	256
$B_c(= B_X)$	4	B_{IO}	8–64
Clock frequency	1 GHz	N_{GBL}	256
C	10 fF	C_{BL}	230 fF

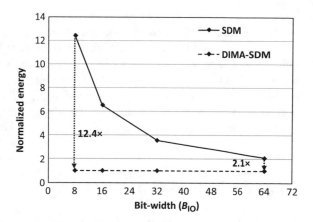

Fig. 5.19 Normalized energy based on models (5.30), (5.31) with $\Delta V_{BL} = 75\,mV$ and $125\,mV$ (obtained from Sect. 5.7.2) for SDM and DIMA-SDM, respectively

5.8 Simulation Results

In this section, SDM is studied in the context of handwritten digit recognition. Monte Carlo circuit (HSPICE) simulations in 65 nm CMOS process technology are employed to validate the models in (5.26) and (5.27). These models are employed in system simulations to estimate the output bad pixel ratio (B_o). Energy and throughput benefits are demonstrated via circuit simulations and the energy/throughput models (5.28)–(5.31).

5.8.1 System Configuration

Nine 16×16 binary shapes of numbers from 1 to 9 are employed to generate p and d as shown in Fig. 5.21a, b. For each of the nine patterns, 225 noisy copies with input bad pixel ratio $B_i = 0.25$ are generated by randomly flipping 25% of the bits. Those images form the training dataset S_t of size $255 \times 9 = 2295$ and were written into the SDM in the auto-associative mode ($p = d$).

In hetero-associative mode, during the training phase, p and d were assigned images corresponding to consecutive numbers, e.g., if p was assigned image corresponding to 4 then d is assigned the image corresponding to 5. Thus, during the READ operation, the SDM retrieves the image corresponding to the number that was one greater than the input number, as shown in Fig. 5.21b.

After the training, 100 contaminated copies of each pattern (for a total of 900 inputs) with $B_i = 0.15,\ 0.25$, and 0.3 were generated and provided as the address p for classification. Four READ iterations of the auto- and hetero-associative memory are performed. The error immunity against faulty hardware increases with larger R in (5.13), as each data vector d is distributed across greater number of physical addresses. On the other hand, R needs to be small enough not to create excessive intersection between physical addresses for different images. To balance that trade-off, we set R for WRITE and READ operations are set to 79 and 82, respectively.

The block size $I/M = 512$ was chosen to balance the read out delay and area efficiency of memory. The value of $B_{IO} = 64$ is chosen as its typical value of J/B_{IO} ranges from 4 to 32 in conventional SRAM architectures [46], in order to permit the maximum bandwidth. In this specific application, $M = 4$ and $S_H \approx 0.1I$, which will be used in the rest of this section. The design parameters (other than B_{IO} and M) used in the simulations are summarized in Table 5.4.

5.8.2 Model Validation

Monte Carlo circuit simulations shows that $\sigma_{cell}/\Delta V_{BL} = 6.5\%$ and that the analog comparator has an input offset $\sigma_{comp} = 18\,\mathrm{mV}$. The behavioral models of the entire analog signal processing chain from the bitcell to the final output V_{sum} were validated as shown in Fig. 5.20, in which the results of Monte Carlo circuit simulations are compared with those from system simulations that employed the behavioral models (5.26) and (5.27). Nine different combinations of p and a_i (with $J = 8$) are chosen as inputs. Figure 5.20 indicates that the maximum modeling error was 4.5% of the dynamic range of V_{sum}. That level of accuracy is sufficient for system performance estimation, as it is much less than the smallest non-zero Hamming distance.

5.8.3 System Performance

The output bad-pixel ratio $B_o[n]$ in the nth READ iteration is computed as:

Fig. 5.20 V_{sum} from circuit simulations and system simulations with behavioral models (5.26) and (5.27) with $J = 8$ and $\Delta V_{BL} = 50\,\mathrm{mV}$

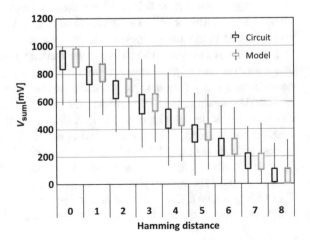

$$B_o[n] = \frac{1}{900J} \sum_{k=1}^{900} H_k[n] \tag{5.32}$$

where $H_k[n]$ is the Hamming distance between the SDM output $\mathbf{y}_k[n]$ at time index n and the ideal output for the kth input image.

Figure 5.21c shows that the conventional SDM, the DIMA-SDM (without HBD) and the DIMA-SDM all converge to achieve a B_o of less than 2% for $n \geq 3$ when $B_i \leq 25\%$. Similar results were observed for the hetero-associative mode as well. Furthermore, SDM and DIMA-SDM were found to achieve $B_o[n]$ that were within $< 5\%$ from each other for $n \leq 3$ (except the B_o of DIMA-SDM was slightly worse), for all three values of B_i. The B_o of DIMA-SDM was higher than SDM by only 0.4% for $n = 4$ and $B_i = 25\%$ indicating that the non-ideal behavior of DIMA-SDM is successfully compensated by the inherent noise immunity of SDM and the associative mode of operation. The B_o degradation of the DIMA-SDM can be reduced by increasing the number of blocks M with large I so that more averaging can occur.

5.8.4 Delay and Energy Savings

The conventional SRAM read access and FR require two clock cycles, and the data transfer from the LDB to the GDB also requires two cycles. The proposed DIMA-SDM achieves 3.1× smaller delay over SDM as shown in Fig. 5.18 due to high bandwidth of DIMA-SDM with $M = 4$.

The various components of the energy models in (5.30) and (5.31) are measured via HSPICE simulations. To do so, the parasitic capacitance of BL ($C_{BL} = 230\,\text{fF}$) is extracted from the layout of an SRAM bitcell. These energy components are a function of the BL swing ΔV_{BL}. The intrinsic robustness of SDM and the associative mode of operation enable a lower value of ΔV_{BL} to be employed as compared to a typical value in standard SRAM, thereby resulting in even greater energy savings.

Figure 5.22 shows the trend of B_o with ΔV_{BL} scaling, and the $B_o > 2\%$ when $\Delta V_{BL} < 75$ and $125\,\text{mV}$ in the conventional SDM and DIMA-SDM, respectively. Thus, energy-optimal ΔV_{BL}s were applied to both conventional SDM and DIMA-SDM in order to obtain the energy breakdowns in Fig. 5.23. The figure shows that DIMA-SDM achieves approximately 2.1× reduced energy as compared to SDM.

5.9 Conclusions

In this chapter, the versatility of DIMA was demonstrated by its use in two applications: CNN and SDM. More specifically, the SDM provides great potential to address the stochastic and unreliable behavior of nanoscale fabrics due to its

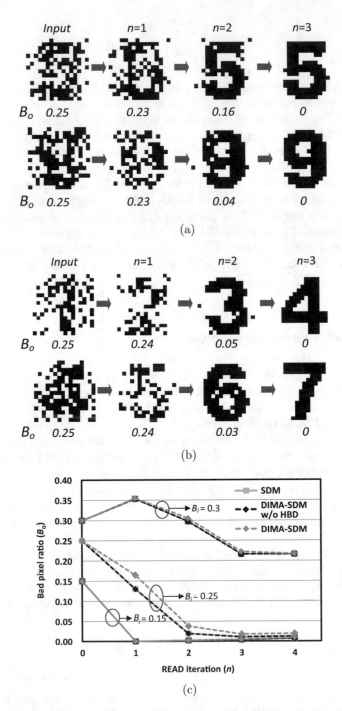

Fig. 5.21 Behavior of SDM and DIMA-SDM (with and without HBD) with $M = 4$ in: (**a**) auto-associative mode, (**b**) hetero-associative mode, and (**c**) the output bad pixel ratio $B_o[n]$ in the auto-associative mode

Fig. 5.22 Energy vs. B_0 trade-off with $n = 4$, where $\Delta V_{BL} = 25$–125 mV for SDM and 75–175 mV for DIMA-SDM

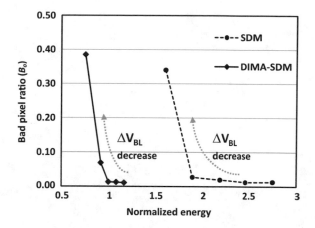

Fig. 5.23 Energy breakdown for a single READ operation with $B_{IO} = 64$, and $\Delta V_{BL} = 75$ mV for SDM and 125 mV for DIMA-SDM

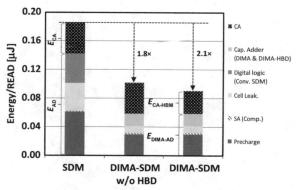

inherent robustness from ensemble decision-making process. However, conventional digital architectures fail to exploit the error tolerance for throughput and energy benefits. This chapter proved that DIMA can be a possible solution to such memory-intensive inference algorithms/applications. It was also demonstrated that the energy and throughput benefits can be further improved by co-optimizing the algorithm as shown in the HBD for SDM and error-aware re-training for CNN. The benefits of proposed architectures are expected to increase with data volume and input data size by saving the data movement costs more. Architecture extensions based on emerging memory topologies are a potential direction of future work.

Chapter 6
PROMISE: A DIMA-Based Accelerator

Previous chapters have shown that DIMA is highly effective in reducing the energy-delay product of decision-making in ML algorithms. However, DIMA's benefits have been demonstrated only for a limited set of functions. That raises the question: *Can DIMA be made programmable without losing much of its energy and throughput benefits over their digital counterparts?* Answering that question is complicated because DIMA relies heavily on highly area-constrained (array pitch-matched) low-swing analog computations, and therefore introduction of programmability into such architectures is a major challenge.

This chapter presents a DIMA-based accelerator called PROMISE [98] for diverse ML algorithms, which tackles all those challenges. PROMISE realizes a high level of programmability without noticeably losing the efficiency of mixed-signal accelerators for specific ML algorithms. PROMISE exposes instruction set mechanisms that allow software control over energy-vs-accuracy tradeoffs, and supports compilation of high-level languages down to the hardware.

6.1 Background

This section presents an analysis of various ML inference algorithms that identifies commonalities in their data-flow. This analysis will show that a DIMA based accelerator is a suitable platform on which to to bulid an ISA (Sect. 6.2) and a compiler (Sect. 6.3) for ML algorithms.

© Springer Nature Switzerland AG 2020
M. Kang et al., *Deep In-memory Architectures for Machine Learning*,
https://doi.org/10.1007/978-3-030-35971-3_6

6.1.1   ML Algorithms

ML algorithms involve repeated vector distance (VD) computations denoted by $D(W_j, X)$ between the N-dimensional input vector X and the weight vector W as depicted in [99]. Commonly used VD computations include the dot product, L1 distance (Manhattan distance), L2 distance (Euclidean distance), and Hamming distance for the ML algorithms including the support vector machine (SVM), template matching, deep neural network (DNN), k-nearest neighbor (k-NN), and matched filter as listed in Fig. 2.1b.

Those ML algorithms have the following three data-flow properties in common: **(P1)** A single VD is obtained by first computing N element-wise scalar distances (SDs) $(d(w[j][i], x[i]))$ and then performing an aggregation step such as a sum or average to generate the final scalar VD $D(W, X) = \sum_{i=1}^{N} d(w[j][i], x[i])$; **(P2)** The VD between a single query vector X and multiple (say N_o) weight vectors W_js ($j = 1, 2, \ldots N_o$) needs to be computed; **(P3)** The VD goes through a simple decision function $f()$, such as sigmoid or ReLU, to generate the decision y_j. Especially, the VD computation tends to dominate the execution time and energy consumption of ML algorithms for practical problems.

6.1.2   Mixed-Signal ML Accelerator

After BLs are pre-charged for a read cycle, DIMA simultaneously asserts B_w WLs (Fig. 2.3b) to develop a voltage drop (ΔV_{BL}) proportional to a binary weighted sum of B_w bits in the corresponding column [29]. That first FR step in DIMA is referred to as in **(S1)** analog Read (aREAD). aREAD seamlessly converts digital values stored in memory into analog values for subsequent analog computation stages.

Following **(S1)**, three subsequent analog processing stages: **(S2)** analog scalar distance (aSD) implementing scalar distance computations right next to the bitcell array; **(S3)** analog vector distance (aVD), performing the aggregation $(\sum_{i=1}^{N}$ in Fig. 2.1b) by simply charge-sharing all the analog outputs from aSD blocks in one shot; and **(S4)** analog-to-digtal conversion (ADC) and threshold (TH), converting the analog output of aVD into a digital word and subsequently generating a final decision from the digital word based on a given decision function $f()$ in Fig. 2.1b. Note that the aSD stage can support scalar comparison, multiplication, subtraction, addition, and absolute computation in bitcell pitch-matched analog circuitry [29, 41], while the ADC and TH stages consume a negligible portion of the total energy, as they operate infrequently (once after ≥ 128 aSD operations).

Previous DIMA ICs support only limited reconfigurability [44]. Furthermore, the absence of an instruction set limits the use to a short sequence of operations with a single computation kernel, a single memory bank, and fixed parameters such as vector length.

6.2 DIMA Instruction Set Architecture

This section presents the PROMISE architecture as a substrate for exploring ISA, discusses various challenges in developing ISAs, and then proposes an ISA for a programmable mixed-signal accelerator.

6.2.1 The PROMISE Architecture

Single Bank Architecture PROMISE is built on the DIMA core in [32] (Fig. 6.1a). Along with **(S1)** aREAD, **(S2)** aSD, **(S3)** aVD, and **(S4)** ADC and TH, described in Sect. 6.1, PROMISE comprises X-REG and CTRL to transform CM into a programmable mixed-signal accelerator.

A PROMISE bank consists of 256 (= N_{COL}) columns. An 8-bit (= B_w) word is distributed across four consecutive rows that constitute a *word row* and two neighboring columns that store a 4-bit MSB and a 4-bit LSB to enhance linearity through a sub-ranged read technique [44]. That is, aREAD can read out a 128-element vector of digital values and seamlessly convert it to corresponding analog values. Furthermore, aREAD can simultaneously perform element-wise addition or subtraction with X, a 128-element vector that represents the input operand for inference in Fig. 2.1b. **aSD** and **aVD** are architected to perform operations on 128 analog values at a time. **ADC** consists of eight 8-bit ADCs that operate in parallel and convert arround 57 million analog values to digital values per second. Note that the aVD output of each bank is digitized by those ADCs to prevent the accumulation of noise from analog operations over the iterations. That digitization also enables a multiple-bank architecture, wherein reliable data transfers between banks are required. **TH** implements non-linear operations such as sigmoid via piece-wise linear approximation [77] not only to compute the decision functions $f()$ in Fig. 2.1b but also to aggregate intermediate computed values when the vector length N is larger than 128.

Finally, **X-REG** is a digital block similar to a vector register file in a SIMD processor, and holds eight 128-element vectors representing eight X values. **CTRL** is a controller for generating enable signals for the aforementioned components based on a given *instruction* and for making the DIMA function as a programmable mixed-signal accelerator.

Multiple Bank Architecture PROMISE can be extended to a multi-bank structure (Fig. 6.1b), which has multiple (up to eight in this work) PAGEs, each of which includes four banks. Thus, long (> 128) vectors can be distributed across multiple banks for parallel processing. PROMISE does not have the scalability limitation of analog operations, as the partial results from each bank are always converted to digital values by the ADCs. Then, the partial sums can be aggregated across different banks via a cross-bank rail, similar to the H-Tree in [26]. The data transfer

(a)

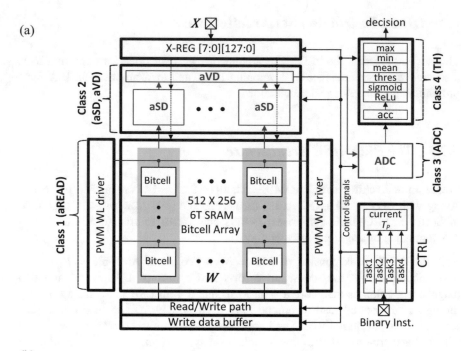

(b)

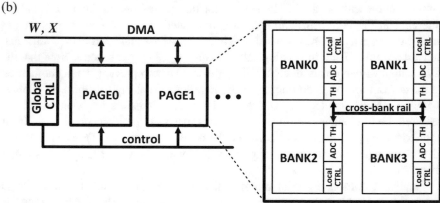

Fig. 6.1 The PROMISE architecture to enable programmability and scalability: (**a**) a single-bank architecture with $N_{ROW} = 512$, $N_{COL} = 256$, with analog processing marked in red, and (**b**) a multi-bank architecture

of 8-bit ADC output from one bank to the other through the cross-bank rail takes only 0.5 pJ with activity factor of 0.5 (obtained from post-layout simulations). That cost is negligible (< 1%) as a aREAD consumes up to 131 pJ/bank as listed in Table 6.2. The control of the multi-bank architecture by the instructions is discussed in Sect. 6.2.3.

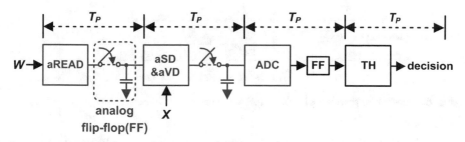

Fig. 6.2 Analog pipeline in PROMISE with operating frequency $= 1/T_P$; and the red area is in the analog domain

Analog Pipeline Architecture To achieve higher throughput than the original DIMA [44], PROMISE adopts analog pipelining as shown in Fig. 6.2. It requires analog flip-flops (FFs) to hold the output of each stage. The first pipelined stage (aREAD) needs N_{COL} analog FFs whereas the second pipelined stage, consisting of aSD and aVD, needs only one analog FF to store the aggregated scalar output. This architecture reuses a pre-existing capacitor, which is a part of the analog multiplier [44] in the aSD block, as an analog FF for the aREAD stage in order to minimize the hardware overhead.

6.2.2   Challenges in Mixed-Signal Accelerator Design

The combination of algorithmic diversity and mixed-signal operations in PROMISE creates many challenges that should be considered when exploring ISA design choices. To support a broad range of ML algorithms, each stage needs to support diverse programmable operations. However, there are some challenges in supporting (too) many diverse programmable operations and they can significantly degrade both throughput and accuracy.

Higher Impact of Delay Variation across Operations Analog operations often exhibit a wide range of delays, depending on the operation types [44], while each stage must accommodate the worst-case delay out of all the operations for the stage. Furthermore, as all the stages of a pipelined mixed-signal accelerator need to operate at the same clock period T_P, T_P needs to accommodate the worst-case delay across all the stages, i.e., $T_P = \max(T_{S1\_max}, T_{S2\_max}, T_{S3\_max}, T_{S4\_max})$, as illustrated in Fig. 6.3. That is, support for a larger number of programmable operations causes longer idle times for some stages. As a result, up to $2\times$ throughput degradation was observed when designing for the range of operations supported in this work.

Furthermore, such long idle times negatively affects accuracy at the algorithm level because analog values are typically stored on (area-constrained) capacitors and thus degrade over time because of various leakage mechanisms. The problem

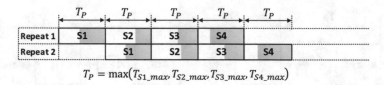

$$T_P = \max(T_{S1\_max}, T_{S2\_max}, T_{S3\_max}, T_{S4\_max})$$

Fig. 6.3 Four-stage processing in PROMISE with operational diversity per stage

is especially bad for the S1 stage (aREAD), as each BL is subject to the leakage contributions from all the bitcells in the column (up to 0.6%/ns).

Limitations on Sequence of Operations There are a few challenges specific to analog processing, which further limits the number of possible programmable operations. Specifically, a chain of analog processing stages imposes intrinsic sequentiality of operations.

Furthermore, two consecutive stages need to be physically closely placed to avoid substantial degradation in analog voltage from one stage to the next. Lastly, a large capacitor ratio between consecutive stages (input-output 20:1 to maintain the voltage drop < 5%) is required to transfer the signal via a charge-transfer mechanism without an additional analog buffer.

6.2.3 PROMISE Instruction Set

This section presents a suitable ISA for a programmable mixed-signal accelerator, considering the above constraints.

Instruction Format PROMISE uses a wide-word macro instruction format, which is referred to as Task. Akin to a very-large instruction word (VLIW) ISAs, a single Task consists of multiple operations, *except that the operations are sequential and not parallel as in VLIW architectures*. As depicted in Fig. 6.4a, the four Class fields specify four operations for four pipelined stages of PROMISE, while the three other fields, OP_PARAM, RPT_NUM, and MULTI_BANK, configure all or specific Class operations. More specific descriptions of the seven fields are as follows.

Operating Parameter Field OP_PARAM comprises 33 bits and configures operating parameters of Class operations in a given Task, facilitating *flexible programmability*. As shown in Fig. 6.4b, W_ADDR specifies a CM address for a Class-1 operation. X_ADDR1 and X_ADDR2 designate X-REG addresses for Class-1 and Class-2, respectively. SWING controls BL swing ΔV_{BL}, e.g., 111 allows 30 mV/LSB, whereas 001 allows 5 mV/LSB. This parameter is a key knob for controlling the trade-off between energy and accuracy under software control; Sect. 6.5 evaluates this accuracy-energy trade-off for further energy savings. Refer to Fig. 6.4c for descriptions of the remaining parameters and their assignments of bit fields.

(a) **Task**

OP_PARAM (28 bits)	RPT_NUM (7 bits)	MULTIBANK (2 bits)	Class-1 (3 bits)	Class-2 (4 bits)	Class-3 (1 bits)	Class-4 (3 bits)

(b)

Contents	Bits	Description
SWING	[27:25]	ΔV_{BL} swing code – 000: min (5 mV/LSB), 111: max (30 mV/LSB)
ACC_NUM	[24:23]	# of operands to be accumulated for accumulate opcode in Class-4
W_ADDR	[22:14]	bitcell array address of W in Class-1
X_ADDR1	[13:11]	bitcell array address of X in Class-1
X_ADDR2	[10:8]	X-REG array address of X in Class-2
X_PRD	[7:6]	X_ADDR1 & 2 circulate from 0 to "X_PRD - 1"
DES	[5:4]	Class-4 output destination - 00: ACC input, 01: output buffer, 10: X-REG, 11:Write data buffer
THRES_VAL	[3:0]	Thresholding reference value for threshold opcode in Class-4

(c)

Class	Operation [operand]	Bit length	OP CODE	Option
1	none	3 bits (OPCODE)	000	X_PRD: X_ADDR1 circulates from 0 to "X_PRD-1"
	write[W_ADDR]		001	
	read[W_ADDR]		010	
	aREAD[W_ADDR]		011	
	aSUBT[W_ADDR, X_ADDR1]		100	
	aADD[W_ADDR, X_ADDR1]		101	
2	none	4 bits (OPCODE + aVD)	000	aVD bit: 0: no aggregation 1: aggregation X_PRD: X_ADDR2 circulates from 0 to "X_PRD-1"
	compare		001	
	absolute		010	
	square		011	
	sign_mult[X_ADDR2]		100	
	unsign_mult[X_ADDR2]		101	
3	none	1 bit (OPCODE)	0	
	ADC		1	
4	accumulation	3 bits (OPCODE)	000	DES, ACC_NUM
	mean		001	DES
	threshold		010	DES, THRES_VAL
	max		011	DES
	min		100	DES
	sigmoid		101	DES
	ReLu		111	DES

Fig. 6.4 PROMISE instruction set: (**a**) instruction format, (**b**) operation parameters (OP_PARAM), and (**c**) operations in each Class

Class Fields Class-1 is composed of 3-bit opcode and defines six possible memory operations. READ, WRITE, or aREAD makes CM perform a digital read, digital write, or analog read operation to a compute-memory address specified by OP_PARAM (W_ADDR). aADD or aSUB fuses an analog read and an element-wise analog addition or subtraction into a single operation in which two vector operands come from compute-memory addresses specified by OP_PARAM (W_ADDR and X_ADDR1).

Class-2 consists of 4-bit opcode and specifies a composition of one of six possible aSD operations with one of two possible aVD operations. Specifically, aSD operating on a computed value from Class-1 supports three unary operations (compare, absolute, and square) and two binary operations (sign_mult and unsign_mult) with the second operand coming from an X-REG address specified by OP_PARAM (X_ADDR2). aVD specifies whether an aggregation should be performed or not after an aSD operation.

Class-3 and Class-4 comprise 1- and 3-bit opcode. Class-3 controls whether or not an ADC should be performed, and Class-4 specifies one of seven possible TH operations. The seven possible TH operations are as follows: accumulation, mean, threshold, max, min, sigmoid, and ReLu.

In summary, Class-1, Class-2 and Class-3 define a distance computation, $D(W_j, X)$ which is shown in Fig. 2.1b in the analog domain, while Class-4 specifies $f(D(W_j, X))$ in the digital domain.

Loop Control Field RPT_NUM comprises 7 bits and specifies how many times the Task should be executed to process multiple W_js. The CM and X-REG addresses (W_ADDR, X_ADDR1, and X_ADDR2) are incremented sequentially every iteration. Although, that is unconventional for a modern RISC architecture, it is a natural choice for typical ML inference algorithms, which iterate sequentially through data W_js in memory for the computation $D(W_j, X)$.

Multiple Bank Control Field MULTI_BANK comprises 2 bits and specifies the number of banks used to distribute long (> 128) vectors for parallel processing. The intermediate results from banks 1, 2,..., and $2^{MULTI\_BANK} - 1$ are transferred to bank 0 through the cross-bank rail in Fig. 6.1b to be aggregated by a digital TH accumulation operation. The long vector needs to be distributed to the same row of each bank to support the parallel processing across multiple banks. The instruction is shared by 2^{MULTI_BANK} banks as those banks process the same operation. One can transfer the output of a Class-4 operation to the X-REG of a specific bank in any PAGE by defining *DES_ADD* in OP_PARAM.

Extension to Large-Scale Applications PROMISE is well suited for 128-dimensional vector processing. In can process longer vectors (> 128) by repeating the 128 dimensional vector processing sequentially by setting RPT_NUM = (W.size() / 128) and other parameters. A word-row of CM stores 128 words, and thus two word-rows are used to store a 256-dimension vector. Two consecutive iterations complete a vector processing. The addresses W_ADDR and X_ADDR1 (or X_ADDR2) are incremented as the Task iterates to process the next 128 words. However, the X is reused to compute the distances to many Ws as explained in Sect. 6.1. Thus, the X_ADDR1 and X_ADDR2 circulate from 0 to X_PRD-1. For example, X_PRD = 2 to process a 256 element vector X.

6.2.4 Algorithm Mapping and Compiler Needs

An ML algorithm sometimes requires several Tasks for different distance metrics. Consider the example of template matching with L1 distance kernel to find the closest 512-pixel image out of 127 candidate images (W_j) to input query image (X) by processing the data across four banks in parallel. Template matching is mathematically defined as:

$$j_{opt} = \arg\min_j \sum_{i=1}^{512} |x[i]w[j, i]| \tag{6.1}$$

The corresponding Task instruction consists of RPT_NUM = 127, specifying the number of candidate images; MULTI_BANK = 4, to distribute 512 pixels into four banks (128 pixels per bank) for parallel processing; Class-1 aSUBT, to perform element-wise subtraction of X with W_j; Class-2 absolute with aggregation; and Class-3 ADC followed by a digital-domain Class-4 min to compute $f() = \arg\min_j$.

Although each Class offers a limited number of operations (6, 12, 2, and 7, respectively), PROMISE can perform more than 1000 compositions of operations for a given X value. Furthermore, the order of Task instructions in a complex ML inference algorithm and the accuracy setting of each Task through the SWING parameter significantly affect the accuracy at the algorithm level, exploding the solution space for code generation. Hence, it is inefficient and inherently sub-optimal to manually generate code for a complex ML inference algorithm. This, in turn, gives a compiler an opportunity to generate and optimize code that implements a given ML inference algorithm, which is topic explored in the next section.

6.3 Compiler

This section describes PROMISE compiler design requirements in for the translation of an ML algorithm described in a high-level language into the PROMISE ISA, and how the compiler addresses the programmability challenges mentioned in Sect. 6.2.2.

6.3.1 Goals

Hardware Abstraction The compiler intermediate representation (IR) should abstract away low-level details of the hardware, so that front ends don't have to be concerned about hardware details like `Class-1` vs. `Class-2` vector operations, or the specifics of the `OP_PARAM` parameters. At the same time, the IR should enable compilers to perform optimizations, and should capture information that is essential for generating efficient code on PROMISE. It should be in a manner similar to that used by mid-level compiler IRs to abstract away details like finite register files and different register classes from front ends, while enabling sophisticated register allocation algorithms to manage these details [100].

Accuracy-Energy Tradeoff Although the PROMISE ISA allows the compiler to use the `SWING` parameter as a knob for tuning ΔV_{BL} to exploit the trade-off between energy and accuracy, it is difficult to go from a high-level description of "accuracy" that an application programmer understands in the context of the algorithm to hardware-specific parameters such as voltage swings. That is especially true for applications that have several smaller computations to offload (for example, DNNs can offload each layers' computation) to a hardware accelerator. It is even more difficult to reason about how the error in a single computation would affect the overall accuracy of an application. Towards that end, the PROMISE compiler must provide a compiler optimization to determine the optimal `SWING` parameter for each `Task`, starting with some application-level specification of accuracy.

Hardware-Specific Optimizations Furthermore, for applications for which the data do not fit into PROMISE memory, the compiler needs to find an efficient data-flow pattern based on the size of the data arrays.

Easily Extensiblility to ML Domain-Specific Languages Domain-specific languages (DSLs) and libraries for ML are evolving fast: there is already a wide range of popular DSLs, such as Torch, Theano, Tensorflow, MXNet, Keras, and others [101–105], implemented on top of dynamic programming languages such as Python, Julia, R, Scala, Perl, and others. Thus, it would be desirable for the PROMISE compiler to be easily extendable to new DSLs. To achieve that, it should provided with a language-neutral IR as an interface for the compiler/programmer.

6.3.2 AbstractTask and PROMISE Compiler IR

An `AbstractTask` is an abstraction of a PROMISE `Task` described in Sect. 6.2.3. `AbstractTask` is based on our observation that a vector operation can be either a `Class-1` (addition/subtraction) or a `Class-2` (signed/unsigned multiplication) operation, but that distinction is relevant only for late-stage code generation, not front ends and other compiler optimizations. `AbstractTask` is

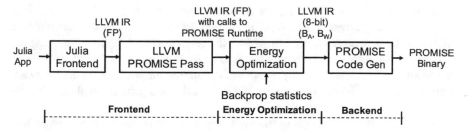

Fig. 6.5 PROMISE compiler Pipeline. LLVM(FP) implies that data arrays are in floating-point. "8-bit" implies fixed-point data arrays

also oblivious to hardware-specific parameters, such as the number of elements in a vector (i.e., the length of the bitcell array), the size of the bitcell array, etc.

6.3.3 Code Generation

Figure 6.5 shows the PROMISE compiler pipeline for Julia. There are three parts: (1) a front end to map Julia applications to the IR, (2) energy optimizations on the IR, and (3) a back end to translate the IR into the PROMISE ISA. The IR, energy optimizations, and back end are all designed to be independent of the source-level language, to make it easily extendable to other languages or DSLs.

6.3.4 Energy Optimization

PROMISE allows the source-level programmer to express the additional error called the *mismatch probability* (p_m) [39] that the users can tolerate when running their model. As deterministic errors can be tolerated easily through retraining the parameters in ML algorithms, the PROMISE compiler focuses on spatial random errors across bitcells arising from process variations. Formally, p_m is the upper bound on the difference between the classification accuracy of an algorithm running on PROMISE ($p_{PROMISE}$) and the classification accuracy of the ML model (p_{model}), i.e.,

$$p_{model} - p_{PROMISE} \leq p_m \qquad (6.2)$$

The energy optimization in the compiler pipeline in Fig. 6.5 takes the mismatch probability p_m from the program and determines the `swing` field of each `AbstractTask` in the application that would ensure that the error tolerance is met. Mapping of a high-level parameter like p_m directly to a suitable swing voltage is difficult, and is even more challenging for algorithms such as neural networks that have multiple Tasks. The compiler solves that problem by breaking it down into two parts, taking advantage of prior work [39] for the first part: (a) determining a minimum bit precision required to achieve the given mismatch probability, using the results of [39] (note that the mismatch probability is an algorithmic property and is hardware-independent); and (b) mapping the required bit precision to the hardware swing voltage, which is a property of PROMISE. the two parts are explained briefly below starting with the second one.

To achieve B-bit precision in the final output, the error introduced must be less than $1/2^{B+1}$. The major source of error in PROMISE due to lowering of the swing voltage in aREAD operations (see Sect. 6.2). The output of aREAD follows a normal distribution $\hat{W} \sim \mathcal{N}(W, \sigma_W^2)$, where $\sigma_W = |W| \cdot f(\text{SWING})$ and $f(\text{SWING})$ is a function of the SWING parameter and ranges from 0.08 to 0.75. The SWING parameter and $f(\text{SWING})$ are inversely proportional, and hence the σ_W is minimized with a higher SWING parameter. After the aggregation of N such vector elements through charge-sharing, the standard deviation of the aggregated value (σ_{agg}) of output is $\sigma_W/\sqrt{N}$. Since W is in range $[-1, 1]$, $|W| = 1$ for all values to maximize σ_W and σ_{agg}. For a confidence level of 99%, the B-bit precision at the output of aggregation is dictated by:

$$2.6\sigma_{\text{agg}} = 2.6\frac{f(\text{SWING})}{\sqrt{N}} < \frac{1}{2^{B+1}} \qquad (6.3)$$

Given a required p_m (6.3) can be used to compute the minimum swing voltage. To achieve (a), Sakr's model from [39] is leveraged to analyze the quantization (floating-point to fixed-point conversion) tolerance of neural networks and give a relationship between p_m and the bit precisions used for the activations and weights (B_A and B_w) of a neural network model.

The analysis bounds the mismatch probability $p_m = p_{\text{fl}} - p_{\text{fp}}$ by,

$$p_m \leq \Delta_A^2 E_A + \Delta_W^2 E_W \qquad (6.4)$$

where E_A and E_W are statistics of the model obtained while training the model, and $\Delta_A = 2^{-(B_A-1)}$, $\Delta_W = 2^{-(B_w-1)}$.

In the context of PROMISE, Sakr's model is used to calculate the bit precision B_x for X in each AbstractTask given the mismatch probability p_m for the model (i.e., sequence of AbstractTasks), and given the weight precision $B_w = 7$ (since the PROMISE bitcell array uses 8-bits to store a value, including one sign bit).

6.4 Validation Methodology

This section describes our methodology for validating PROMISE's energy, delay, and accuracy benefits as well as the benefits of the compiler-generated code. Fig. 6.6 summarizes the validation methodology. Specifically, (a) energy, delay,

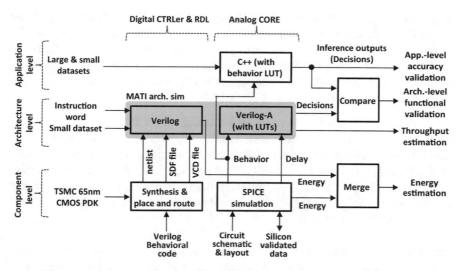

Fig. 6.6 PROMISE validation methodologies at component, architecture and application levels

and behavioral models of PROMISE components in TSMC 65 nm GP process including analog non-idealities were developed; (b) component-level analog models (in Verilog-A) with Verilog model of the digital CTRL to ensure correct functionality and estimate accuracy over small datasets were incorporated; and (c) PROMISE C++ model with component level behavioral models for verifying accuracy over large datasets of compiler -enerated code was developed.

Component-Level Models The entire mixed-signal chain was post-layout simulated in SPICE to obtain the energy and delay numbers listed in Table 6.2. The total energy and delay for the mixed-signal blocks were compared with the measured results reported in [44] and the differences were found to be within 10 and 9%, respectively. The deterministic errors of the analog components were extracted from measurements in the form of look-up tables (LUTs). On the other hand, the spatial random error across bitcells due to the process variation was extracted from Monte-Carlo SPICE simulations to obtain a statistically sufficient number of samples. Behavioral models incorporating these non-ideal analog effects and delays were then incorporated into component-level Verilog-A models for analog blocks. Verilog models of all the digital components including the CTRL and the TH blocks, were developed and synthesized with the same library via the Synopsys Design Compiler.

Architecture-Level Validation The Verilog and Verilog-A models described in Sect. 6.4 were integrated to obtain a cycle and a functionally accurate PROMISE Verilog model. The digital blocks were verified by generating the correct control signals at the right time in the presence of post-layout parasitics when presented with the appropriate PROMISE instruction word for small datasets.

Application-Level Validation A functional PROMISE C++ model incorporating the LUT-based analog behavioral models described earlier in Sect. 6.4 was also developed. This C++ model was run on large datasets to obtain PROMISE's application-level accuracy. The compiler generated code was verified with the Verilog models of the digital components along with the Verilog-A models to evaluate the energy and accuracy of PROMISE.

Benchmarks The commonly employed ML algorithms listed in Table 6.1 were mapped to PROMISE. As PROMISE is programmable, it employs 8-bit data to cover diverse applications and algorithm, as shown in many other implementations [1, 58, 59, 106]. For application, level energy optimization analysis, three variants of DNN of different levels of complexity were chosen to demonstrate the architecture. These benchmarks give us diversity in complexity and allows exploration of the energy-vs-accuracy trade-offs.

Baseline Architectures It is well known that ASICs are couple of orders-of-magnitude more energy, and delay-efficient than general-purpose processors (CPU/GPU) [111]. Therefore, the following four ASICs were chosen as comparison baselines for a conservative evaluation of PROMISE. (a) **CM**: The programmability overhead is estimated via a comparison with DIMA. (b) **State-of-the-art**: PROMISE is compared to recent prior silicon IC prototypes [7, 58], that implement algorithms similar to those of PROMISE. The DNN is compared with [7], and k-NN and template matching with L1 and L2 distances are compared with [58]. (c) **CONV-8b**: A baseline digital architecture was built (Fig. 6.7) with the 8-b fixed point computational logic synthesized for the specific algorithm + conventional SRAM. (d) **CONV-OPT**: This is the same as CONV-8b but with a minimum bit precision required per benchmark.

Even though prior art exists for some benchmarks, many of them do not have a relevant previous ASIC. Furthermore, configurations such as for process technology and on-chip memory capacity are not perfectly identical to those of PROMISE. In order to perform a conservative comparison, the CONV-8b/CONV-OPT (CONV) is chosen to operate at maximum speed while restricting the number of SRAM banks employed to be the same as in PROMISE. The SRAM fetches $N_{\text{COL}}/(L B_{\text{w}})$ words per single read access of a bank. Therefore, CONV operates with a maximum achievable throughput of:

$$f_{\text{conv}} = \left(\frac{N_{\text{COL}}/L}{B_{\text{w}}} \right) \left(\frac{1}{T_{\text{SRAM}}} \right) \tag{6.5}$$

CONV (Fig. 6.7) consists of computation logic synthesized for each specific benchmark; therefore. it incurs only the energy costs of that specific benchmark, and additional routing, dataflow, and control energy are neglected. In addition, CONV-OPT has the minimum precision for each benchmark, thereby making it the most conservative baseline to compare with PROMISE in terms of accuracy, energy, and throughput.

Table 6.1 Benchmarks for PROMISE simulations [62–64]

Algorithm	Application	Database	Data size (N)	Number of categories	Problem size	# of AbstractTasks	AbstractTask	W	X	Comments	Opt. swing ($p_m = 1\%$)
DNN (multi-layer perceptron)	Hand-written character recognition	MNIST	(8-bit) 22×23 10	10 categories, 60,000 training samples, 10,000 test samples	4	vecOp: multiplication redOp: sum	Weights	Test samples	5-layer DNN with nodes as follows: 784-512-256-128-10	3,2,3,3	
Matched filtering	Event (gun-shot) Detection	Gun-shot mono sound	(8-bit) 256 512 1024	2	100 test vectors	1	vecOp: multiplication redOp: sum	Filter weights	Test samples		1
Template matching (w/ L1 & L2)	Face recognition	MIT-CBCL	(8-bit) 16×16 22×23 32×33	64	256 candidates	1	vecOp: subtraction redOp: L1 – absolute L2 – square	Candidate faces	Test samples	Nearest candidate based on either L1 or L2 distance	2
Linear SVM	Face detection	MIT-CBCL	(8-bit) 16×16	2	2 categories, 2000 training samples, 858 test samples	1	vecOp: multiplication redOp: sum	Weights	Test samples	Face data converted into a vector, linear SVM applied on it	6

(continued)

Table 6.1 (continued)

Algorithm (w/ L1 & L2)	Application	Database	Data size (N)	Number of categories	Problem size	# of AbstractTasks	AbstractTask	W	X	Comments	Opt. swing (p_m =1%)
k-NN (w/ L1 & L2)	Hand-written character recognition	MNIST	(8-bit) 16×16 22×23 32×33	10	10 categories, 54210 training samples, 200 test samples	1	vecOp: subtraction redOp: L1 – absolute L2 – square	Training samples	Test samples	Sorting is done in external processor after processing on PROMISE	1
Feature extraction (PCA)	Face detection	MIT-CBCL	(8-bit) 16×16	–	2000 samples	1	vecOp: multiplication redOp: sum	Weights	Samples	Four features used for face detection based on PCA	–
Linear regression	Modeling linear predictor	Synthetic data	(8-bit) 2 dim.	–	8192 samples	4	vecOp: None redOp: AT1, AT2 – mean AT3 – square AT4 – sign_mult	AT1: U AT2: V AT3: U AT4: U	AT4: V	2-D linear regression : $slope = \frac{\Sigma uv - \Sigma \bar{u}\bar{v}}{\Sigma u^2 - \Sigma \bar{u}^2}$ $y\text{-}intercept = \bar{v}$ $-slope \cdot \bar{u}$	–

Fig. 6.7 Single bank of CONV-8b with $L = 4$, $N_{\text{ROW}} = 512$, and $N_{\text{COL}} = 256$, where an SRAM communicates with an algorithm-specific synthesized digital processor over a pipelined interface

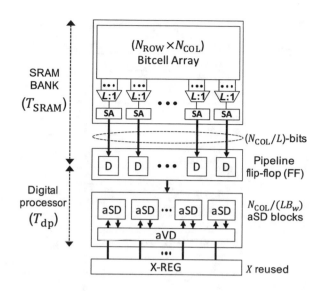

6.5 Evaluation

This section presents the evaluation of PROMISE. The energy and delay for each operation were estimated using the methodology presented in Sect. 6.4. The gains of the compiler-based energy optimization are presented and compared with the maximum (unoptimized) swing voltages, and then compare PROMISE against the baselines described in Sect. 6.4.

PROMISE executes one 128-element vector operation per bank within T_P, i.e., its throughput in terms of "number of OPs per bank per unit time" can be expressed by $f_{\text{PROMISE}} = 128/T_P$ per bank. The energy consumption can be divided as shown:

$$E_{\text{PROMISE}} = \sum_{i=1}^{4} E_{\text{Class},i} + E_{\text{LEAK}} + E_{\text{CTRL}} \qquad (6.6)$$

where $E_{\text{Class},i}$ is the energy consumed by the Class,i instruction, and E_{CTRL} and E_{LEAK} are the CTRL block and leakage energies, respectively. Table 6.2 shows the energy consumed for each operation at SWING = 111.

6.5.1 Effectiveness of Compiler

Code Generation The benchmarks for evaluation are listed in Table 6.1. They use a wide variety of combinations of operations in different Classes. Encoding of these diverse algorithms by hand or using a library is not feasible. Coding them in Julia, and using the compiler to generate a PROMISE ISA, was both more

Table 6.2 Energy & delay per operation (1 cycle = 1 ns)

Class	Operation	Delay(# of cycles)	Energy/bank (pJ)
1	`write`	2	73
	`read`	2	33
	`aREAD`	5	61
	`aSUBT`	7	103
	`aADD`	7	103
2	`compare`	6	5
	`absolute`	6	12
	`square`	8	38
	`sign_mult`	14	16
	`unsign_mult`	14	16
3	`ADC`	138	6
4	`accumulation`	4	≈ 0
	`mean`	3	≈ 0
	`threshold`	2	≈ 0
	`max`	4	≈ 0
	`min`	4	≈ 0
	`sigmoid`	3	≈ 0
	`ReLu`	3	≈ 0
Leakage energy per cycle (1 ns)			0.6
CTRL energy per cycle (1 ns)			5.4

efficient and error free. Moreover, it is suboptimal to find the `SWING` parameter for benchmarks that use more than one `Task`. For example, for the DNN benchmark with three hidden layers, the number of `SWING` combinations is $8^4 = 4096$. Inter `Task` compiler analysis is required to find the optimal `SWING` parameter for such algorithms, as shown later. Lastly, the compiler also handles different vector sizes for these benchmarks.

Energy Optimization The energy benefits on PROMISE were evaluated by choosing the optimal `swing` values obtained via compiler-directed energy optimization. Figure 6.8 shows the energy benefits of energy-optimized code generated for PROMISE. The figure contains energy estimates of PROMISE for the benchmarks under test for two cases: (1) full precision for which all `Tasks` use maximum `SWING`, and (2) optimized for which `Tasks` use the `SWING` is set by the energy optimization pass. The degradation in accuracy was limited to 1% ($p_m = 1\%$). Feature extraction and linear regression are omitted from this evaluation as they are not classification kernels, and mismatch probability is defined for classification algorithms only.

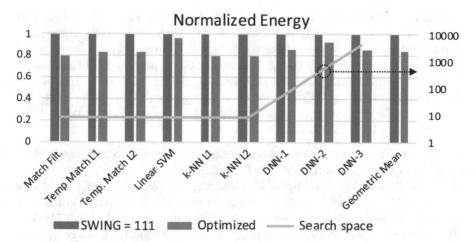

Fig. 6.8 Energy gains by compiler-directed optimization. DNN-(1, 2, 3) are trained on MNIST dataset. Their structures are: DNN-1(784-128-10), DNN-2(784-256-128-10), and DNN-3(784-512-256-128-10)

The first six benchmarks in Fig. 6.8 compile down to a single `AbstractTask` in the PROMISE compiler IR, and the optimal `swing` was obtained by doing a sweep over all eight values of `swing`. The last three benchmarks, DNN-1, DNN-2, and DNN-3, are variants of the multilayer perceptron shown in Table 6.1. The three DNNs that have 3, 4, and 5 layers translate to 2, 3, and 4 `AbstractTasks`, respectively. The search space for the optimal `swing` increases exponentially with the increase of each layer. Energy optimization analysis was used to obtain the `swing` values for these DNNs. The optimal `swing` for the `AbstractTasks` is (3, 6) for DNN-1, (5, 7, 7) for DNN-2, and (3, 3, 4, 6) for DNN-3. The maximum energy savings come from the lower layers of each DNN, which are wider and also more tolerant to imprecision. Overall, the benefits of the optimization range from 4% for Linear SVM to about 20% for the two k-NN versions (geometric mean: 15%).

6.5.2   Performance and Energy

Comparison with State-of-the-Art   The IC in [58] implements k-NN accelerator with L1 and L2-distances in a 14 nm FinFET process, where an 8-bit, 128-dimension X is processed with 128 W_js. The k-NN accelerator achieves 3.37 (3.84) nJ/decision with 21.5M (20.3M) decisions/s with L1 (L2) distance. PROMISE achieves 18 (22.9) nJ/decision with 1.12M (0.98M) decisions/s with L1 (L2) distance for the same benchmark with single bank. Though PROMISE achieves lower energy efficiency and throughput, it is implemented in a 65 nm process (vs. 14 nm FinFET in [58]). When the energy and delay numbers in [58] are scaled to a 65 nm process based on ITRS roadmap [112], PROMISE achieves 4.1× (3.7×) smaller energy and 3.1× (3.4×) lower throughput, achieving 1.3× (1.1×) energy-delay product (EDP) reduction with L1 (L2) distance.

The DNN accelerator [7] was implemented in a 28 nm process for an 8-bit, 5-layer DNN with a network size of 784-256-256-256-10. The accelerator employs total a 1 MB SRAM (to test up to a 16-bit case), zero-skipping, and the RAZOR technique [17], and achieves 0.57 μJ/decision and 28K decisions/s. On the other hand, PROMISE enables an 8-bit, 5-layer DNN with a size of 784-512-256-128-10 in 36 banks (= 576 KB). The network sizes are not identical, but comparable (PROMISE's network is slightly larger, requiring 69% higher number of coefficients W_js and MAC operations). PROMISE achieves 0.49 μJ/decision and 558K decisions/s, achieving 1.15× energy saving and 19.9× throughput improvement with a 22× EDP reduction, though PROMISE was implemented in a 65 nm process (vs. 28 nm in [7]).

Comparison with CONV Figure 6.9a shows that PROMISE provides a speed-up of 1.4 − 3.4× compared to CONV-OPT across the benchmarks. PROMISE's speed-up is the least for linear regression because it needs to reaccess the same SRAM data for every Task because analog data cannot be stored (because of leakage) whereas CONV stores the data in a local register (pipeline FF in Fig. 6.7) and reuses it. Figure 6.9b shows that PROMISE achieves a 3.4 − 5.5× energy savings compared to CONV-OPT, leading to EDP improvements of 4.7 − 12.6× compared to CONV-OPT. The key reason for PROMISE's energy efficiency is that aREAD (Class-1) and aSD/aVD (Class-2) are executed with a low-voltage swing mixed-signal computation block (See Fig. 6.10).

The programability overhead of PROMISE is minimal. Estimates show the concepts introduced in this chapter actually can improve over DIMA. Results show that PROMISE achieves up to 1.9× speed-up over DIMA due to the analog pipeline in spite of its operational diversity. In spite of the increased complexity of CTRL to support the programmability, PROMISE was found to achieve 5.5% energy savings over CM due to reduced leakage, as PROMISE can go to sleep mode quicker after completing the given Tasks due to the throughput gain.

(a)

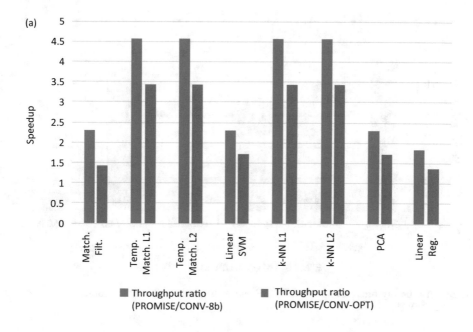

(b)

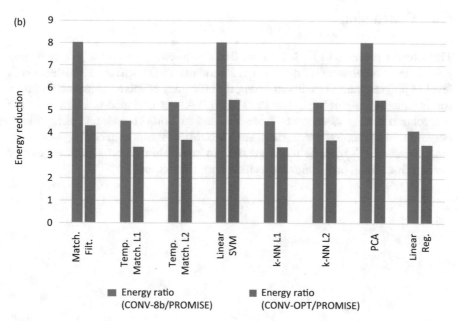

Fig. 6.9 PROMISE gains (with SWING = 111) compared to CONV in terms of: (**a**) speed-up, and (**b**) energy savings

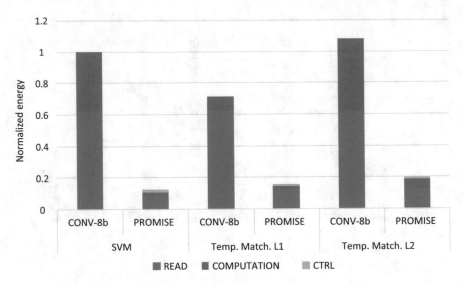

Fig. 6.10 Energy breakdowns of PROMISE and conventional architecture (normalized to SVM with CONV-8b)

6.6   Conclusion

This chapter presented PROMISE, the first end-to-end design of a programmable mixed-signal accelerator for diverse ML algorithms. PROMISE accomplishes a high level of programmability without losing the efficiency of mixed-signal accelerators for specific ML algorithms. The PROMISE ISA was designed to allow software control over energy-vs-accuracy trade-offs, and supports compilation of high-level languages like Julia down to the hardware. PROMISE shows better energy efficiency than digital ASICs, despite much greater programmability, and significant energy savings from small programmer-specified error tolerances.

Chapter 7
Future Prospects

This book has described a unique architectural concept referred to as the *deep in-memory architecture* (DIMA), for implementing data-centric workloads found in emerging applications such as in health care, social networks, smart infrastructure, and surveillance/monitoring among others. DIMA addresses the high energy and latency costs of data movement between the processor and memory by embedding mixed-signal computations deeply into the periphery of the memory core. The DIMA concept [29] and its 65 nm CMOS IC realizations [32, 33, 52] have targeted aggressive energy reductions via SNR vs. energy trade-off to demonstrate up to $100\times$ reduction in the decision EDP. DIMA's versatility was shown by mapping a diverse set of inference algorithms, ranging from the simple (SVM, template matching, k-NN, matched filter) to the complex (CNN and SDM), to a DIMA-based programmable accelerator architecture.

Since the publication of the DIMA concept paper [29], a community of in-memory computing researchers has emerged that is exploring various approaches to realize in-memory architectures for machine learning. As a result, in-memory architectures are an exciting and growing area of research today, and many new in-memory design approaches remain to be discovered. Some of the more promising research directions on the topic of in-memory computing are outlined below.

Scalable In-memory Architectures In-memory architectures such as DIMA are, in essence, a super-efficient matrix-vector multiply (MVM) engines. Such architectures in fact shine the most when processing high-dimensional data. Though an MVM is a key computational kernel in deep networks, there exists no good method for mapping complex networks onto an in-memory architecture. Challenges include: (1) the mismatch between the physical and algorithmic array sizes; (2) the need to realize residual (non-DIMA) computations; (3) the difficulty of exploiting reuse opportunities; and (4) the need for compiler or auto-tuner techniques to map high-level application metrics (e.g., accuracy of decisions) to DIMA parameters. Chapter 6 discusses a first step in addressing those challenges via the design of an ISA for DIMAs along with compiler support, but much work remains to be done

© Springer Nature Switzerland AG 2020
M. Kang et al., *Deep In-memory Architectures for Machine Learning*,
https://doi.org/10.1007/978-3-030-35971-3_7

to demonstrate their effectiveness via real-life prototypes capable of implementing a diversity of large networks for real-life users. Making DIMA-based hardware platforms augmented with software libraries will be critical for enabling DIMA's widespread use. Some work in that direction has already begun [113].

In-memory Architectures in Emerging Memory Technologies Materials and device researchers have been prolific in proposing novel memory technologies beyond SRAM, DRAM, and flash. Technologies such as PCM, RRAM, MRAM, have been proposed and others are being developed. Though many of the current in-memory architectures have focused on SRAMs, there is a huge opportunity to explore such architectures in non-CMOS memory technologies. Some work in that direction has already been done, such as extension of DIMA into flash [114] and MRAM [115], as well as development of cross-bar arrays for RRAMs [116]. The introduction of a new device fundamentally alters the in-memory design space, invariably generating a rich and productive area of research. In fact, the ability of DIMA and other in-memory techniques to average out the impact of circuit/device noise prevalent in emerging memory technologies, makes them well-matched to the attributes of memory devices. Naturally, that compatibility raises the exciting prospect of device-(in-memory) architecture co-design that requires device researchers to collaborate closely with circuit designers and architects to devise new device concepts that are informed by the unique needs of in-memory architectures.

Error Compensation Techniques for In-memory Architectures In-memory architectures operate under a fundamental energy vs. SNR trade-off to achieve large reductions in decision EDP. Thus far, in-memory architectures have operated well within the error-tolerance envelope intrinsic to machine learning algorithms in order to minimize the impact of their low-SNR operation on system accuracy. Much greater reductions in EDP are feasible if error compensation techniques are harnessed to push the limits of the energy efficiency of in-memory architectures. Doing so would require judicious use of traditional error-control methods such as coding, as well as methods such as Shannon-inspired statistical error compensation (SEC) [117] to in-memory architectures. The use of on-chip training [118] is one way of pushing the SNR lower without compromising on decision accuracy. Indeed, the true limits of any new device technology can only be found when such algorithmic techniques are combined with various architectural and circuit level methods to combat the effect of low-SNR operation.

The research opportunities described above encompasses systems, algorithms, architectures, circuits and devices. In this sense, in-memory computing is a full-stack technology that requires system designers, architects, and circuit and device researchers to collaborate. It is hoped that this book has provided the reader with a glimpse of this unique technology and its promise.

Correction to: Deep In-memory Architectures for Machine Learning

Correction to:
M. Kang et al., *Deep In-memory Architectures*
for Machine Learning,
https://doi.org/10.1007/978-3-030-35971-3

This book was inadvertently published with an incorrect affiliation of the author "Mingu Kang". The affiliation details have been corrected now.

The updated version of the book can be found at
https://doi.org/10.1007/978-3-030-35971-3

References

1. D. Silver, A. Huang, C.J. Maddison, A. Guez, L. Sifre, G. Van Den Driessche, J. Schrittwieser, I. Antonoglou, V. Panneershelvam, M. Lanctot et al., Mastering the game of Go with deep neural networks and tree search. Nature **529**(7587), 484–489 (2016)
2. A. Krizhevsky, I. Sutskever, G.E. Hinton, Imagenet classification with deep convolutional neural networks, in *Advances in Neural Information Processing Systems (NIPS)* (2012), pp. 1097–1105
3. J. Baliga, R.W. Ayre, K. Hinton, R.S. Tucker, Green cloud computing: balancing energy in processing, storage, and transport. Proc. IEEE **99**(1), 149–167 (2011)
4. Y.-H. Chen, T. Krishna, J.S. Emer, V. Sze, Eyeriss: an energy-efficient reconfigurable accelerator for deep convolutional neural networks. IEEE J. Solid State Circuits **52**(1), 127–138 (2017)
5. B. Moons, R. Uytterhoeven, W. Dehaene, M. Verhelst, Envision: a 0.26-to-10TOPS/W subword-parallel dynamic-voltage-accuracy-frequency-scalable convolutional neural network processor in 28nm FDSOI, in *IEEE International Solid-State Circuits Conference (ISSCC)* (2017), pp. 246–247
6. D. Shin, J. Lee, J. Lee, H.-J. Yoo, DNPU: an 8.1 TOPS/W reconfigurable CNN-RNN processor for general-purpose deep neural networks, in *IEEE International Solid-State Circuits Conference (ISSCC)* (2017), pp. 240–241
7. P.N. Whatmough, S.K. Lee, H. Lee, S. Rama, D. Brooks, G.-Y. Wei, A 28nm SoC with a 1.2 GHz 568nJ/prediction sparse deep-neural-network engine with >0.1 timing error rate tolerance for IoT applications, in *IEEE International Solid-State Circuits Conference (ISSCC)* (2017), pp. 242–243
8. J. Lee, C. Kim, S. Kang, D. Shin, S. Kim, H.-J. Yoo, UNPU: a 50.6 TOPS/W unified deep neural network accelerator with 1b-to-16b fully-variable weight bit-precision, in *IEEE International Solid-State Circuits Conference (ISSCC)* (2018), pp. 218–220
9. T. Chen, Z. Du, N. Sun, J. Wang, C. Wu, Y. Chen, O. Temam, Diannao: a small-footprint high-throughput accelerator for ubiquitous machine-learning, in *ACM Sigplan Notices*, vol. 49, no. 4 (2014), pp. 269–284
10. M. Horowitz, Computing's energy problem (and what we can do about it), in *IEEE International Solid-State Circuits Conference (ISSCC)* (2014), pp. 10–14
11. A. Firoozshahian, *Smart Memories: A Reconfigurable Memory System Architecture* (ProQuest, Ann Arbor, 2009)
12. K. Mai, T. Paaske, N. Jayasena, R. Ho, W.J. Dally, M. Horowitz, Smart memories: a modular reconfigurable architecture, in *ACM SIGARCH Computer Architecture News*, vol. 28 (ACM, New York, 2000), pp. 161–171

© Springer Nature Switzerland AG 2020

M. Kang et al., *Deep In-memory Architectures for Machine Learning*,

https://doi.org/10.1007/978-3-030-35971-3

13. K. Mai, R. Ho, E. Alon, D. Liu, Y. Kim, D. Patil, M.A. Horowitz, Architecture and circuit techniques for a 1.1-GHz 16-kb reconfigurable memory in 0.18-μm CMOS. IEEE J. Solid State Circuits **40**(1), 261–275 (2005)

14. D. Patterson, T. Anderson, N. Cardwell, R. Fromm, K. Keeton, C. Kozyrakis, R. Thomas, K. Yelick, Intelligent RAM (IRAM): chips that remember and compute, in *IEEE International Solid-State Circuits Conference (ISSCC)* (IEEE, Piscataway, 1997), pp. 224–225

15. Y.-H. Chen, J. Emer, V. Sze, Eyeriss: a spatial architecture for energy-efficient dataflow for convolutional neural networks, in *International Symposium on Computer Architecture (ISCA)* (2016), pp. 367–379

16. M. Price, J. Glass, A.P. Chandrakasan, A scalable speech recognizer with deep-neural-network acoustic models and voice-activated power gating, in *IEEE International Solid-State Circuits Conference (ISSCC)* (2017), pp. 244–245

17. D. Ernst, N.S. Kim, S. Das, S. Pant, R. Rao, T. Pham, C. Ziesler, D. Blaauw, T. Austin, K. Flautner et al., RAZOR: a low-power pipeline based on circuit-level timing speculation, in *IEEE/ACM International Symposium on Microarchitecture (MICRO)* (2003), pp. 7–18

18. J.P. Kulkarni, K. Kim, K. Roy, A 160 mV robust Schmitt trigger based subthreshold SRAM. IEEE J. Solid State Circuits **42**(10), 2303–2313 (2007)

19. B. Zhai, D. Blaauw, D. Sylvester, S. Hanson, A sub-200mV 6T SRAM in 0.13 μm CMOS, in *IEEE International Solid-State Circuits Conference (ISSCC)* (IEEE, Piscataway, 2007), pp. 332–606

20. F. Frustaci, M. Khayatzadeh, D. Blaauw, D. Sylvester, M. Alioto, SRAM for error-tolerant applications with dynamic energy-quality management in 28 nm CMOS. IEEE J. Solid State Circuits **50**(5), 1310–1323 (2015)

21. F. Frustaci, D. Blaauw, D. Sylvester, M. Alioto, Approximate SRAMs with dynamic energy-quality management. IEEE Trans. Very Large Scale Integr. Syst. **24**(6), 2128–2141 (2016)

22. K. Bong, S. Choi, C. Kim, S. Kang, Y. Kim, H.-J. Yoo, A 0.62 mW ultra-low-power convolutional-neural-network face-recognition processor and a CIS integrated with always-on Haar-like face detector, in *IEEE International Solid-State Circuits Conference (ISSCC)* (2017), pp. 248–249

23. H.J. Mattausch, T. Gyohten, Y. Soda, T. Koide, Compact associative-memory architecture with fully parallel search capability for the minimum Hamming distance. IEEE J. Solid State Circuits **37**(2), 218–227 (2002)

24. Y. Oike, M. Ikeda, K. Asada, A high-speed and low-voltage associative co-processor with exact Hamming/Manhattan-distance estimation using word-parallel and hierarchical search architecture. IEEE J. Solid State Circuits **39**(8), 1383–1387 (2004)

25. R. Genov, G. Cauwenberghs, Kerneltron: support vector "machine" in silicon. IEEE Trans. Neural Netw. **14**(5), 1426–1434 (2003)

26. S. Aga, S. Jeloka, A. Subramaniyan, S. Narayanasamy, D. Blaauw, R. Das, Compute caches, in *IEEE International Symposium on High Performance Computer Architecture (HPCA)* (2017), pp. 481–492

27. J. Wang, X. Wang, C. Eckert, A. Subramaniyan, R. Das, D. Blaauw, D. Sylvester, A compute sram with bit-serial integer/floating-point operations for programmable in-memory vector acceleration, in *IEEE International Solid-State Circuits Conference (ISSCC)* (IEEE, Piscataway, 2019), pp. 224–226

28. D. Bankman, L. Yang, B. Moons, M. Verhelst, B. Murmann, An always-on 3.8uJ/86% CIFAR-10 mixed-signal binary CNN processor with all memory on chip in 28-nm CMOS. IEEE J. Solid State Circuits **54**(1), 158–172 (2018)

29. M. Kang, M.-S. Keel, N.R. Shanbhag, S. Eilert, K. Curewitz, An energy-efficient VLSI architecture for pattern recognition via deep embedding of computation in SRAM, in *IEEE International Conference on Acoustics, Speech and Signal Processing (ICASSP)* (2014), pp. 8326–8330

30. N. Shanbhag, M. Kang, M.-S. Keel, Compute memory. Issued July 4 2017, US Patent 9,697,877 B2

31. J. Zhang, Z. Wang, N. Verma, In-memory computation of a machine-learning classifier in a standard 6T SRAM array. IEEE J. Solid State Circuits **52**(4), 915–924 (2017)
32. M. Kang, S.K. Gonugondla, A. Patil, N.R. Shanbhag, A multi-functional in-memory inference processor using a standard 6T SRAM array. IEEE J. Solid State Circuits **53**(2), 642–655 (2018)
33. M. Kang, S.K. Gonugondla, S. Lim, N.R. Shanbhag, A 19.4-nJ/decision, 364-K decisions/s, in-memory random forest multi-class inference accelerator. IEEE J. Solid State Circuits **53**(7), 2126–2135 (2018)
34. A. Biswas, A.P. Chandrakasan, Conv-RAM: an energy-efficient SRAM with embedded convolution computation for low-power CNN-based machine learning applications, in *IEEE International Solid-State Circuits Conference (ISSCC)* (2018), pp. 488–490
35. W.-H. Chen, K.-X. Li, W.-Y. Lin, K.-H. Hsu, P.-Y. Li, C.-H. Yang, C.-X. Xue, E.-Y. Yang, Y.-K. Chen, Y.-S. Chang et al., A 65nm 1Mb nonvolatile computing-in-memory ReRAM macro with sub-16ns multiply-and-accumulate for binary DNN AI edge processors, in *IEEE International Solid-State Circuits Conference (ISSCC)* (2018), pp. 494–496
36. W.-S. Khwa, J.-J. Chen, J.-F. Li, X. Si, E.-Y. Yang, X. Sun, R. Liu, P.-Y. Chen, Q. Li, S. Yu et al., A 65nm 4kb algorithm-dependent computing-in-memory SRAM unit-macro with 2.3 ns and 55.8 TOPS/W fully parallel product-sum operation for binary DNN edge processors, in *IEEE International Solid-State Circuits Conference (ISSCC)* (2018), pp. 496–498
37. T.F. Wu, H. Li, P.-C. Huang, A. Rahimi, J.M. Rabaey, H.-S.P. Wong, M.M. Shulaker, S. Mitra, Brain-inspired computing exploiting carbon nanotube FETs and resistive RAM: hyperdimensional computing case study, in *IEEE International Solid-State Circuits Conference (ISSCC)* (2018), pp. 492–494
38. I. Hubara, M. Courbariaux, D. Soudry, R. El-Yaniv, Y. Bengio, Binarized neural networks, in *Advances in Neural Information Processing Systems (NIPS)* (2016), pp. 4107–4115
39. C. Sakr, Y. Kim, N. Shanbhag, Analytical guarantees on numerical precision of deep neural networks, in *International Conference on Machine Learning (ICML)* (2017), pp. 3007–3016
40. B. Moons, K. Goetschalckx, N. Van Berckelae, M. Verhelst, Minimum energy quantized neural networks, in *Asilomar Conference on Signals, Systems and Computer* (2017)
41. M. Kang, S.K. Gonugondla, M.-S. Keel, N.R. Shanbhag, An energy-efficient memory-based high-throughput VLSI architecture for convolutional networks, in *IEEE International Conference on Acoustics, Speech and Signal Processing (ICASSP)* (2015)
42. M. Kang, S.K. Gonugondla, N.R. Shanbhag, A 19.4 nJ/decision 364K decisions/s in-memory random forest classifier in 6T SRAM array, in *IEEE European Solid-State Circuits Conference (ESSCIRC)* (2017), pp. 263–266
43. M. Kang, E.P. Kim, M.-S. Keel, N. R. Shanbhag, Energy-efficient and high throughput sparse distributed memory architecture, in *IEEE International Symposium on Circuits and Systems (ISCAS)* (2015), pp. 2505–2508
44. M. Kang, S. Gonugondla, A. Patil, N. Shanbhag, A 481pJ/decision 3.4M decision/s multifunctional deep in-memory inference processor using standard 6T SRAM array. arXiv:1610.07501 (preprint, 2016)
45. J. Backus, Can programming be liberated from the von Neumann style?: a functional style and its algebra of programs. Commun. ACM **21**(8), 613–641 (1978)
46. M. Yamaoka et al., A 300-MHz 25-μA/Mb-leakage on-chip SRAM module featuring process-variation immunity and low-leakage-active mode for mobile-phone application processor. IEEE J. Solid State Circuits **40**(1), 186–194 (2005)
47. J. Zhang, Z. Wang, N. Verma, A machine-learning classifier implemented in a standard 6T SRAM array, in *IEEE Symposium on VLSI Circuits (VLSI Circuits)* (2016), pp. 1–2
48. R.G. Dreslinski, M. Wieckowski, D. Blaauw, D. Sylvester, T. Mudge, Near-threshold computing: reclaiming Moore's law through energy efficient integrated circuits. Proc. IEEE **98**(2), 253–266 (2010)
49. K.J. Kuhn, Reducing variation in advanced logic technologies: approaches to process and design for manufacturability of nanoscale CMOS, in *IEEE International Electron Devices Meeting (IEDM)* (IEEE, Piscataway, 2007), pp. 471–474

50. D. Bankman, B. Murmann, An 8-bit, 16 input, 3.2 pJ/op switched-capacitor dot product circuit in 28-nm FDSOI CMOS, in *IEEE Asian Solid-State Circuits Conference (A-SSCC)* (2016), pp. 21–24

51. S. Assefa, S. Shank, W. Green, M. Khater, E. Kiewra, C. Reinholm, S. Kamlapurkar, A. Rylyakov, C. Schow, F. Horst et al., A 90nm CMOS integrated nano-photonics technology for 25Gbps WDM optical communications applications, in *IEEE International Electron Devices Meeting (IEDM)*, 2012, pp. 33–38

52. S.K. Gonugondla, M. Kang, N.R. Shanbhag, A variation-tolerant in-memory machine learning classifier via on-chip training. *IEEE J. Solid State Circuits* **53**(11), 3163–3173 (2018)

53. Z.Wang, K.H. Lee, N. Verma, Overcoming computational errors in sensing platforms through embedded machine-learning kernels. *IEEE Trans. Very Large Scale Integr.* **23**(8), 1459–1470 (2015)

54. M. Kang, S. Lim, S. Gonugondla, N.R. Shanbhag, An in-memory VLSI architecture for convolutional neural networks. *IEEE J. Emerg. Sel. Top. Circuits Syst.* **8**(3), 494–505 (2018)

55. Y. Kim, M. Kang, L.R. Varshney, N.R. Shanbhag, Generalized water-filling for source-aware energy-efficient SRAMs. *IEEE Trans. Commun.* **66**(10), 4826–4841 (2018)

56. Z. Zhou, B. Pain, E.R. Fossum, CMOS active pixel sensor with on-chip successive approximation analog-to-digital converter. *IEEE Trans. Electron Devices* **44**(10), 1759–1763 (1997)

57. F. Arnaud, F. Boeuf, F. Salvetti, D. Lenoble, F. Wacquant, C. Regnier, P. Morin, N. Emonet, E. Denis, J. Oberlin et al., A functional 0.69 μm^2 embedded 6T-SRAM bit cell for 65nm CMOS platform, in *IEEE Symposium on VLSI Technology (VLSI Technology)* (2003), pp. 65–66.

58. H. Kaul, M.A. Anders, S.K. Mathew, G. Chen, S.K. Satpathy, S.K. Hsu, A. Agarwal, R.K. Krishnamurthy, A 21.5 M-query-vectors/s 3.37 nJ/vector reconfigurable k-nearest-neighbor accelerator with adaptive precision in 14nm tri-gate CMOS, in *IEEE International Solid-State Circuits Conference (ISSCC)* (2016), pp. 260–261

59. S. Gupta, A. Agrawal, K. Gopalakrishnan, P. Narayanan, Deep learning with limited numerical precision, in *International Conference on Machine Learning (ICML)* (2015), pp. 1737–1746

60. S.C. Chung, Circuits and methods of a self-timed high speed SRAM, Nov. 10, 2015. US Patent 9,183,897

61. E. Karl, Y. Wang, Y.-G. Ng, Z. Guo, F. Hamzaoglu, M. Meterelliyoz, J. Keane, U. Bhattacharya, K. Zhang, K. Mistry et al., A 4.6 GHz 162 Mb SRAM design in 22 nm tri-gate CMOS technology with integrated read and write assist circuitry. *IEEE J. Solid State Circuits* **48**(1), 150–158 (2013)

62. Center for biologicaland computational learning (CBCL) at MIT (2000). http://cbcl.mit.edu/software-datasets/index.html

63. Production Crate, Gun shot sounds. http://soundscrate.com/gun-related

64. Y. LeCun, C. Cortes, MNIST handwritten digit database. AT&T Labs (2010). http://yann.lecun.com/exdb/mnist

65. L. Breiman, Random forests. Mach. Learn. **45**(1), 5–32 (2001)

66. T. Kobayashi, K. Nogami, T. Shirotori, Y. Fujimoto, A current-controlled latch sense amplifier and a static power-saving input buffer for low-power architecture. *IEEE J. Solid State Circuits* **76**(5), 863–867 (1993)

67. V.A. Prisacariu, R. Timofte, K. Zimmermann, I. Reid, L. Van Gool, Integrating object detection with 3D tracking towards a better driver assistance system, in *IEEE International Conference on Pattern Recognition (ICPR)* (2010), pp. 3344–3347

68. B. Van Essen, C. Macaraeg, M. Gokhale, R. Prenger, Accelerating a random forest classifier: multi-core, GP-GPU, or FPGA? in *IEEE Annual International Symposium on Field-Programmable Custom Computing Machines (FCCM)* (2012), pp. 232–239

69. T.-W. Chen, Y.-C. Su, K.-Y. Huang, Y.-M. Tsai, S.-Y. Chien, L.-G. Chen, Visual vocabulary processor based on binary tree architecture for real-time object recognition in full-HD resolution. *IEEE Trans. Very Large Scale Integr. Syst.* **20**(12), 2329–2332 (2012)

70. K.J. Lee, G. Kim, J. Park, H.-J. Yoo, A vocabulary forest object matching processor with 2.07 M-vector/s throughput and 13.3 nJ/vector per-vector energy for full-HD 60 fps video object recognition. *IEEE J. Solid State Circuits* **50**(4), 1059–1069 (2015)

71. L. Bottou, F.E. Curtis, J. Nocedal, Optimization methods for large-scale machine learning. arXiv:1606.04838 (preprint, 2016)

72. C. Cortes, V. Vapnik, Support-vector networks. Mach. Learn. **20**(3), 273–297 (1995)

73. C. Sakr, A. Patil, S. Zhang, Y. Kim, N. Shanbhag, Minimum precision requirements for the SVM-SGD learning algorithm, in *IEEE International Conference on Acoustics, Speech and Signal Processing (ICASSP)* (2017), pp. 1138–1142

74. M. Kang, N.R. Shanbhag, In-memory computing architectures for sparse distributed memory. *IEEE Trans. Biomed. Circuits Syst.* **10**(4), 855–863 (2016)

75. Y. LeCun, L. Jackel, L. Bottou, A. Brunot, C. Cortes, J. Denker, H. Drucker, I. Guyon, U. Muller, E. Sackinger et al., Comparison of learning algorithms for handwritten digit recognition, in *International Conference on Artificial Neural Networks*, vol. 60 (1995), pp. 53–60

76. D. Strigl, K. Kofler, S. Podlipnig, Performance and scalability of GPU-based convolutional neural networks, in *IEEE Euromicro International Conference on Parallel, Distributed and Network-Based Processing (PDP)* (2010), pp. 317–324

77. C. Farabet, C. Poulet, J.Y. Han, Y. LeCun, CNP: an FPGA-based processor for convolutional networks, in *IEEE International Conference on Field Programmable Logic and Applications (FPL)* (2009), pp. 32–37

78. K. Simonyan, A. Zisserman, Very deep convolutional networks for large-scale image recognition. arXiv:1409.1556 (preprint, 2014)

79. K. He, X. Zhang, S. Ren, J. Sun, Deep residual learning for image recognition, in *IEEE Conference on Computer Vision and Pattern Recognition (CVPR)* (2016), pp. 770–778

80. W. Rieutort-Louis, T. Moy, Z. Wang, S. Wagner, J.C. Sturm, N. Verma, A large-area image sensing and detection system based on embedded thin-film classifiers. *IEEE J. Solid State Circuits* **51**(1), 281–290 (2016)

81. B. Moons, M. Verhelst, A 0.3-2.6 TOPS/W precision-scalable processor for real-time large-scale ConvNets, in *IEEE Symposium on VLSI Circuits (VLSI Circuits)* (2016), pp. 1–2

82. J.K. Kim, P. Knag, T. Chen, Z. Zhang, A 640M pixel/s 3.65 mW sparse event-driven neuromorphic object recognition processor with on-chip learning, in *IEEE Symposium on VLSI Circuits (VLSI Circuits)* (2015), pp. C50–C51

83. J.M. Cruz-Albrecht, M.W. Yung, N. Srinivasa, Energy-efficient neuron, synapse and STDP integrated circuits. *IEEE Trans. Biomed. Circuits Syst.* **6**(3), 246–256 (2012)

84. S. Brink, S. Nease, P. Hasler, S. Ramakrishnan, R. Wunderlich, A. Basu, B. Degnan, A learning-enabled neuron array IC based upon transistor channel models of biological phenomena. *IEEE Trans. Biomed. Circuits Syst.* **7**(1), 71–81 (2013)

85. S. Ramakrishnan, R. Wunderlich, J. Hasler, S. George, Neuron array with plastic synapses and programmable dendrites. *IEEE Trans. Biomed. Circuits Syst.* **7**(5), 631–642 (2013)

86. V. Garg, R. Shekhar, J.G. Harris, Spiking neuron computation with the time machine. *IEEE Trans. Biomed. Circuits Syst.* **6**(2), 142–155 (2012)

87. P. Kanerva, *Sparse distributed memory* (MIT Press, Cambridge, 1988)

88. P.J. Denning, *Sparse distributed memory*. Research Institute for Advanced Computer Science (NASA Ames Research Center, Mountain View, 1989)

89. E. Lehtonen, J.H. Poikonen, M. Laiho, P. Kanerva, Large-scale memristive associative memories. *IEEE Trans. Very Large Scale Integr. Syst.* **22**(3), 562–574 (2014)

90. M. Lindell et al., Configurable sparse distributed memory hardware implementation, in *IEEE International Symposium on Circuits and Systems (ISCAS)* (1991), 3078–3081

91. J. Saarinen et al., VLSI architectures of sparse distributed memory, in *IEEE International Symposium on Circuits and Systems (ISCAS)* (1991), pp. 3074–3077

92. J.D. Keeler et al., Notes on implementation of sparsely distributed memory, in *NASA Research Institute for Advanced Computer Science* (1986)

93. S.-I. Chien, I.-C. Kim, D.-Y. Kim, Iterative autoassociative memory models for image recalls and pattern classifications, in *IEEE International Joint Conference on Neural Networks (IJCNN)* (1991), pp. 30–35

94. I. Kim et al., High performance PRAM cell scalable to sub-20nm technology with below $4F^2$ cell size, extendable to DRAM applications, in *IEEE Symposium on VLSI Technology (VLSI Technology)* (2010), pp. 203–204

95. S. Aritome, Advanced Flash memory technology and trends for file storage application, in *IEEE International Electron Devices Meeting (IEDM)* (2000), pp. 763–766

96. A. Verma, B. Razavi, Frequency-based measurement of mismatches between small capacitors, in *IEEE Custom Integrated Circuits Conference (CICC)* (2006), pp. 481–484

97. K. Kim, H. Mahmoodi, K. Roy, A low-power SRAM using bit-line charge-recycling technique, in *International Symposium on Low Power Electronics and Design (ISLPED)* (2007), pp. 177–182

98. P. Srivastava, M. Kang, S.K. Gonugondla, S. Lim, J. Choi, V. Adve, N.S. Kim, N. Shanbhag, PROMISE: an end-to-end design of a programmable mixed-signal accelerator for machine-learning algorithms, in *Proceedings of the 45th Annual International Symposium on Computer Architecture* (IEEE Press, Piscataway, 2018), pp. 43–56

99. D. Liu, T. Chen, S. Liu, J. Zhou, S. Zhou, O. Teman, X. Feng, X. Zhou, Y. Chen, PuDianNao: a polyvalent machine learning accelerator, in *ACM SIGARCH Computer Architecture News*, vol. 43, no. 1. (ACM, New York, 2015), pp. 369–381

100. A.V. Aho, M.S. Lam, R. Sethi, J.D. Ullman, *Compilers: Principles, Techniques, and Tools*, 2nd edn. (Addison-Wesley, Boston, 2006)

101. R. Collobert, K. Kavukcuoglu, C. Farabet, Torch7: a Matlab-like environment for machine learning, in *BigLearn, NIPS Workshop* (2011)

102. R. Al-Rfou, G. Alain, A. Almahairi, C. Angermüller, D. Bahdanau, N. Ballas et al., Theano: a Python framework for fast computation of mathematical expressions. CoRR, vol. abs/1605.02688 (2016)

103. M. Abadi, P. Barham, J. Chen, Z. Chen, A. Davis, J. Dean, M. Devin, S. Ghemawat, G. Irving, M. Isard, M. Kudlur, J. Levenberg, R. Monga, S. Moore, D.G. Murray, B. Steiner, P. Tucker, V. Vasudevan, P. Warden, M. Wicke, Y. Yu, X. Zheng, Tensorflow: a system for large-scale machine learning, in *USENIX Symposium on Operating Systems Design and Implementation (OSDI 16)* (2016)

104. T. Chen, M. Li, Y. Li, M. Lin, N. Wang, M. Wang, T. Xiao, B. Xu, C. Zhang, Z. Zhang, MXNet: a flexible and efficient machine learning library for heterogeneous distributed systems. CoRR, vol. abs/1512.01274 (2015)

105. F. Chollet et al. Keras (2015). https://github.com/fchollet/keras

106. B. Murmann, D. Bankman, E. Chai, D. Miyashita, L. Yang, Mixed-signal circuits for embedded machine-learning applications, in *IEEE 49th Asilomar Conference on Signals, Systems and Computers* (2015), pp. 1341–1345

107. Y. LeCun, Y. Bengio, G. Hinton, Deep learning. *Nature* **521**(7553), 436–444 (2015)

108. K.I. Kim, K. Jung, H.J. Kim, Face recognition using kernel principal component analysis. *IEEE Signal Process. Lett.* **9**(2), 40–42 (2002)

109. R. Brunelli, T. Poggio, Face recognition: features versus templates. *IEEE Trans. Pattern Anal. Mach. Intell.* **15**(10), 1042–1052 (1993)

110. D.K. Mellinger, S.W. Martin, R.P. Morrissey, L. Thomas, J.J. Yosco, A method for detecting whistles, moans, and other frequency contour sounds. *J. Acoust. Soc. Am.* **129**(6), 4055–4061 (2011)

111. R. Hameed, W. Qadeer, M. Wachs, O. Azizi, A. Solomatnikov, B.C. Lee, S. Richardson, C. Kozyrakis, M. Horowitz, Understanding sources of inefficiency in general-purpose chips, in *ACM SIGARCH Computer Architecture News*, vol. 38 (ACM, New York, 2010), pp. 37–47

112. ITRS, ITRS Roadmap [Online]. http://www.itrs2.net/

113. H. Jia, Y. Tang, H. Valavi, J. Zhang, N. Verma, A microprocessor implemented in 65nm CMOS with configurable and bit-scalable accelerator for programmable in-memory computing, arXiv:1811.04047 (preprint, 2018)

114. S.K. Gonugondla, M. Kang, Y. Kim, M. Helm, S. Eilert, N. Shanbhag, Energy-efficient deep in-memory architecture for NAND flash memories, in *2018 IEEE International Symposium on Circuits and Systems (ISCAS)* (IEEE, Piscataway, 2018), pp. 1–5
115. A.D. Patil, H. Hua, S. Gonugondla, M. Kang, N.R. Shanbhag, An MRAM-based deep in-memory architecture for deep neural networks, in *2019 IEEE International Symposium on Circuits and Systems (ISCAS)* (IEEE, Piscataway, 2019), pp. 1–5
116. H.D. Lee et al., Integration of $4F^2$ selector-less crossbar array 2Mb ReRAM based on transition metal oxides for high density memory applications, in *IEEE Symposium on VLSI Technology (VLSI Technology)* (2012), pp. 151–152
117. N.R. Shanbhag, N. Verma, Y. Kim, A.D. Patil, L.R. Varshney, Shannon-inspired statistical computing for the nanoscale era. *Proc. IEEE* **107**(1), 90–107 (2018)
118. S.K. Gonugondla, M. Kang, N. Shanbhag, A 42pJ/decision 3.12 TOPS/W robust in-memory machine learning classifier with on-chip training, in *IEEE International Solid-State Circuits Conference (ISSCC)* (2018), pp. 490–492

Index

A

Accelerators
    DNN, 158
    mixed-signal, 140
Address decoder (AD), 121–122, 125–128
Artificial intelligence (AI), 1
Associative memory, 5, 123–124, 136
Auto-associative memory, 123, 124, 133, 136

B

Biomedical devices, 1
Bit error-rate (BER), 16, 17, 84
BL processing (BLP), 9
    absolute difference, 23
    and CBLP operations, 22
    operation, 22
    pitch-matched layouts, 24
    redistribution-based multiplication, 25–56
Brain-inspired computing, 120
See also Sparse distributed memory (SDM)

C

Charge-recycling mixed-signal multiplier, 104, 109–111
Chip-specific spatial process variations, 87
Column muxing, 11, 12, 17, 50, 66
Compiler
    accuracy-energy tradeoff, 148
    design, 147
    DSLs, 148
    effectiveness, 155–157
    hardware-specific optimizations, 148
    intermediate representation (IR), 148

Conventional digital reference architecture (CONV), 97, 98, 152, 158, 159
Convolutional layers (C-layers), 102–104, 106, 114–116
Convolutional neural networks (CNN), 4, 6
    convolutional layers (C-layers), 102–103
    description, 101
    GPU and FPGA-based implementations, 101
    implementation challenges, 103–104
    multi-layer network, 102
CORE output
    accuracy, 56–57
    energy consumption, 58
    energy-delay product, 60
    MD mode, 60
    trade-offs, 59
Counter array (CA), 121–122, 128–129
Counter array using hierarchical binary decision (CA-HBD), 125, 126, 128–129
Cross BL processing (CBLP), 12
    and CBLP (*see* Cross BL processing (CBLP))
    output voltage, 26–27

D

Data-flow
    algorithmic, 8
    DIMA-CNN, 111
    DP, 7
    inference algorithms, 8
    properties, 140
Data reduction and decision, 11

© Springer Nature Switzerland AG 2020
M. Kang et al., *Deep In-memory Architectures for Machine Learning*,
https://doi.org/10.1007/978-3-030-35971-3

Printed in the United States
By Bookmasters